1993

SO-BJJ-402

3 0301 00076023 7

Bioenergetics
2

Bioenergetics 2

DAVID G. NICHOLLS

Department of Biochemistry,
University of Dundee,
Dundee, UK

STUART J. FERGUSON

Department of Biochemistry,
University of Oxford,
Fellow of St Edmund Hall,
Oxford, UK

LIBRARY
College of St. Francis
JOLIET, ILLINOIS

ACADEMIC PRESS

Harcourt Brace Jovanovich, Publishers

London   San Diego   New York
Boston   Sydney   Tokyo   Toronto

ACADEMIC PRESS LIMITED
24–28 Oval Road
LONDON NW1 7DX

United States Edition published by
ACADEMIC PRESS INC.
San Diego, CA 92101

Copyright © 1992 by
ACADEMIC PRESS LIMITED

First edition published 1982

All Rights Reserved
No part of this book may be reproduced in any form by photostat,
microfilm, or by any other means, without prior written
permission from the publishers

This book is printed on acid-free paper

A catalogue record for this book is available from the British Library

ISBN 0–12–518124–8

Typeset by Mathematical Composition Setters Ltd, Salisbury, Wiltshire
Printed and bound in Great Britain by Mackays of Chatham PLC, Chatham, Kent

574.19121
N614

CONTENTS

147002

PREFACE

The first edition of *Bioenergetics* was designed as an introduction to the chemiosmotic theory. It was conceived immediately following the award of the Nobel prize in Chemistry to Peter Mitchell, at a time when most students, and also many active research workers in the field, did not fully appreciate the scope and intellectual beauty of the theory. In the ten years that have followed the first edition the experimental basis of the theory has become even more firmly established. Thus we no longer consider that it is essential to argue the merits of the chemiosmotic theory *versus* 'chemical' or 'conformational' coupling hypotheses, particularly since original champions of alternative theories who are still active researchers, have adopted the chemiosmotic paradigm. It is, however, dangerous to be complacent about a theory, and we do analyze at length, in Chapter 4, quantitative modifications which have been suggested to the basic 'delocalized' proton circuit.

The chemiosmotic theory, although fundamentally simple and relating to the very core of molecular biochemistry, frequently receives too superficial a treatment in general textbooks. One of our purposes in preparing this second edition is to clarify such topics as the thermodynamics of bioenergetic processes and the stoichiometries of energy coupling reactions which cannot be explained in a few lines of text. However, an even more important impetus for the second edition has been the dramatic advance in knowledge of the molecular structure, and hence mechanism, of energy-transducing proteins which has taken place in the last decade. Most dramatic has been the solution at high resolution of the structures of the bacterial photosynthetic reaction centre and bacteriorhodopsin (Chapter 6). Given that light-harvesting 'antenna' proteins are also yielding to direct structural analysis, it is evident that more is known of the structure/function relationships of membrane proteins involved in energy transduction than for any other class of membrane protein.

We have drawn on the advice of many colleagues to whom we offer our thanks. We hope that they will appreciate that we do not name them individually solely because we wish to take responsibility for our personal interpretation of the material (and for any errors). The mark of a dynamic field is that current findings will sooner or later be superseded: where we have anticipated imminent advances or where we recognize that current knowledge is incomplete we have reminded the reader that the text was completed in the latter half of 1991 and thus contains information available up to that point.

As in the first edition, we have attempted to be as concise as possible, since we are aware that a book such as this becomes increasingly inaccessible to the student reader as the length increases. Thanks to the productivity of our colleagues, however, this edition is almost twice the length of the first edition. We have been subjective and selective in our citation of the literature; in general our policy has been to cite one or more recent reviews for each section of the book where there have been significant developments. Where a specific point needed direct literature support, or where the development was so recent that a suitable review was not yet available, we have cited specific research papers. This policy means that selections for the bibliography will not reflect scientific precedence, and we trust that our colleagues will appreciate this.

Writing this book has inevitably distracted us from our families and our colleagues in our laboratories. We thank them for their patience and support. One of us (SJF) would also like to remember here Jim Lloyd who died during 1991. Jim, as a post-doctoral fellow, was a great help to SJF during his early days as a graduate student in bioenergetics.

Finally, as this book was being sent to press, we were saddened to learn of the death of the founder of modern bioenergetics, Peter Mitchell. Peter inspired a whole generation of scientists with the beauty, elegance and above all testability of the chemiosmotic theory. This book, which summarizes contributions from hundreds of 'chemiosmotic' scientists is, in a small way, a testament to the enormous impact which Peter's ideas have had on the development of the field.

D. G. Nicholls
S. J. Ferguson

NOTE TO THE READER

Two points of nomenclature deserve special attention. First, we have used the symbol Δp for protonmotive force in units of mV. In the first edition, as is frequently done elsewhere, we used $\Delta\tilde{\mu}_{H^+}$, but strictly speaking the latter has units of kJ mol^{-1} and so we have adopted Δp in this edition. Second, we have defined throughout the side of a membrane to which protons are pumped as the P (positive) side and the side from which they are pumped as the N (negative) side. This allows a uniform nomenclature and overcomes the confusion that can arise when describing the matrix side of the inner mitochondrial membrane as being on the inside in mitochondria but on the outside in inverted submitochondrial particles. We realise that P is also used by electron microscopists to define the protoplasmic side of a membrane, e.g. the interior surface of a bacterial cytoplasmic membrane and that this is the N side in our convention, but we believe the advantages and increasing use of the P and N nomenclature outweigh any slight chance of the two conventions being confused.

GLOSSARY

Ac	acetate
AcAc	acetoacetate
acetate	ethanoate
acetic acid	ethanoic acid
ADP/O	The number of molecules of ADP phosphorylated to ATP when two electrons are transferred from a substrate through a respiratory chain to reduce one 'O' ($\frac{1}{2}O_2$) (dimensionless).
ADP/2e$^-$	As ADP/O except more general as the final electron acceptor can be other than oxygen (dimensionless).
Bchl	bacteriochlorophyll
bR	bacteriorhodopsin
Bpheo	bacteriopheophytin
BQ	benzoquinone
BQH$_2$	benzoquinol
$[Ca^{2+}]_c$	cytoplasmic calcium concentration
$[Ca^{2+}]_m$	mitochondrial matrix calcium concentration
Chl	chlorophyll
$C_M H^+$	The effective proton conductance of a membrane or a membrane component (dimensions: nmol H$^+$ min^{-1} mg protein^{-1} mV protonmotive force^{-1}).
Cyt	cytochrome
Cyt bc_1 complex	another name for Complex III (ubiquinol–cytochrome c oxidoreductase)
dO/dt	Respiratory rate (dimensions: nmol O min^{-1} mg protein^{-1})

DAD	diaminodurene (or 2,3,5,6 tetramethyl-p-phenylenediamine)
DBMIB	2,5-dibromo-3-methyl-6-isopropylbenzoquinone
DCCD	N,N'-dicyclohexylcarbodiimide (inhibitor of the F_0 sector of ATP synthase)
DCMU	3-(3,4-dichlorophenyl)-1,1-dimethylurea
DCPIP	2,6-dichlorophenolindophenol
E	Redox potential at any specified set of component concentrations (dimensions: mV).
E^0	Standard redox potential (all components) in their standard states, i.e. 1M solutions and 1 atm gases (dimensions: mV).
$E^{0'}$	Standard redox potential except that pH specified, usually pH = 7 (all other components in their standard states, i.e. 1M solutions and 1 atm gases (dimensions: mV).
E_m	Mid-point potential at a defined pH (equivalent to E^0 at pH = 0 because concentrations of oxidised and reduced species cancel and thus their ratio is unity as in E^0 definition) (dimensions: mV).
$E_{m,7}$	Mid-point potential at pH = 7 (also equivalent to $E^{0'}$, because concentrations of oxidised and reduced species cancel and thus their ratio is unity as in $E^{0'}$ definition) (dimensions: mV).
E_h	Actual redox potential at a defined pH (dimensions: mV).
$E_{h,7}$	Actual redox potential at pH = 7 (dimensions: mV).
EP(S)R	Electron paramagnetic (spin) resonance
F	Faraday constant ($= 0.0965$ kJ mol^{-1} mV^{-1}); to convert from mV to kJ mol^{-1}, multiply by F.
$F_1 F_0$	The proton translocating ATP synthase/ATPase (meaning fraction one/fraction oligomycin)
FCCP	carbonyl cyanide p-trifluoromethoxyphenylhydrazone.
Fd	ferredoxin
Fe/S	iron–sulphur centre
ferricyanide	hexacyanoferrate (III)
ferrocyanide	hexacyanoferrate (II)
G	Gibbs (free) energy
H	enthalpy
H^+/ATP	The number of protons translocated through the ATP synthase for the synthesis of one molecule of ATP (dimensionless) (usually refers to ATP synthase alone but in the case of mitochondria may also subsume H^+ movement associated with adenine nucleotide and phosphate translocation).
H^+/O	The number of protons translocated by a respiratory chain during the passage of 2 electrons from substrate to oxygen (dimensionless).
$H^+/2e^-$	As H^+/O except more general as final electron acceptor need not be oxygen (dimensionless).

$h\nu$	The energy in a photon (dimensions: kJ).
J_{H^+}	proton current (dimensions: nmol H^+ min^{-1} mg protein^{-1}).
K	absolute equilibrium constant
K'	apparent equilibrium constant
LH	light harvesting
mV	millivolt
MV^+	reduced methyl viologen
MV^{2+}	oxidized methyl viologen
N-side/N-phase	Negative side of a membrane from which protons are pumped.
O	$\frac{1}{2}O_2$
P/O	As ADP/O
$P/2e^-$	As $ADP/2e^-$
P-side/P-phase	Positive side of a membrane to which protons are pumped.
P_{870} etc.	The primary photochemically active component in a reaction centre.
PC	plastocyanin
Pheo	pheophytin
PMS	phenazinemethosulphate
PQ	plastoquinone
PQH_2	plastoquinol
PSI, PSII	photosystem I, II
q^+/O	The number of charges translocated across a membrane when 2 electrons are transferred from a substrate to oxygen via an electron transport system.
$q^+/2e^-$	As q^+/O except more general so as to specify the movement of electrons through any segment of an electron transport system.
R	The gas constant (8.3 $J\,mol^{-1}K^{-1}$)
S	Entropy
SHAM	salicylhydroxamic acid
SMP	submitochondrial particle
TMPD	N,N,N'N',-tetramethyl-p-phenylenediamine; a redox mediator, especially between ascorbate and cytochrome c
TPP^+	tetraphenyl-phosphonium cation
UQ, UQH_2	ubiquinone, ubiquinol
Γ	mass action ratio
Γ'	apparent mass action ratio
Δp	protonmotive force (dimensions: mV)
$\Delta\psi$	Membrane potential, i.e. the electrical potential difference between two bulk phases separated by a membrane (dimensions: mV).
ΔpH	The pH difference between two bulk phases on either side of a membrane (dimensionless).
ΔE_h	Difference between two redox couples (dimensions: mV).

ΔG	Gibbs energy change at any specified set of reactant and product concentrations (activities).
ΔG^0	Standard Gibbs energy change when all reactants and products are in their standard states (i.e. 1 M for solutes, pure liquid for solvents and 1 atm. for gases (dimensions: $kJ\ mol^{-1}$).
$\Delta G^{0\prime}$	Standard Gibbs energy change except that H^+ concentration is 10^{-7} (i.e. pH = 7) rather than 1 (i.e. pH = 0).
ΔG_p	The phosphorylation potential, i.e. the ΔG for ATP synthesis at any given set of ATP, ADP and Pi concentrations (dimensions: $kJ\ mol^{-1}$).
ΔH	enthalpy change
ΔS	entropy change
$\Delta \tilde{\mu}_{x^+}$	ion electrochemical gradient
/	antiporter
:	symporter

Mitchell sets sail for the Chemiosmotic New World, despite dire warnings that he will be consumed

1 CHEMIOSMOTIC ENERGY TRANSDUCTION

1.1 INTRODUCTION

Since all biochemical reactions involve energy changes, the term 'bioenergetics' could be applied to the whole of biochemistry and much of cell biology. Bioenergetics as a discipline rose to prominence in the 1950s as a highly directed search for the solution to a single problem: the mechanism by which the energy made available by the oxidation of substrates, or by the absorption of light, could be coupled to 'uphill' reactions such as the synthesis of ATP from ADP and Pi, or the accumulation of ions across a membrane. Within this narrow definition of bioenergetics the central concept – the chemiosmotic theory which forms the core of this book – can be considered established and it might therefore be considered that bioenergetics is a quiescent topic, suffering from the very success of research in the period 1965–1980. However, the novel concepts which emerged during this period, particularly that bioenergetics and membrane transport were irretrievably linked processes, have meant that bioenergetic principles are being applied not only to mitochondria, chloroplasts and bacterial transport, but also to the whole spectrum of transport processes occurring across plasma membranes, intracellular organelle membranes and specialized organelles such as neuronal synaptic vesicles. Finally, while energy transduction is considered 'solved' at the phenomenological level, it must be borne in mind that we are only beginning to acquire a molecular knowledge of the mechanism of ion transport. Indeed it is significant that the first edition of this book was written two years after the awarding of a Nobel Prize to Peter Mitchell for his description of chemiosmotic coupling, while the second edition was preceded by a second 'bioenergetic' Nobel Prize to Johann Deisenhofer, Robert Huber and Hartmut Michel for the first precise atomic structure of an energy-transducing molecule − the bacterial reaction centre which converts light energy into a charge separation across a membrane in less than a microsecond. It is the advances in molecular understanding of such mechanisms in the past decade which have made a new edition of this book necessary. Remarkably, more is now known about the structures of membrane proteins involved in photosynthetic energy transduction than about membrane proteins involved in any other function.

1.2 THE CHEMIOSMOTIC THEORY

Although some ATP synthesis is catalysed by soluble enzyme systems, by far the largest proportion is associated with membrane-bound enzyme complexes which are restricted to a particular class of membrane. These 'energy-transducing' membranes are the plasma membrane of simple prokaryotic cells such as bacteria or blue-green algae, the inner membrane of mitochondria and the thylakoid membrane of chloroplasts (Fig. 1.1). These membranes have a related evolutionary origin, since chloroplasts and mitochondria are commonly thought to have evolved from a symbiotic relationship between a primitive, non-respiring eukaryotic cell and an invading prokaryote. Thus the mechanism of ATP synthesis and ion transport associated with these diverse membranes is sufficiently related, despite the differing natures of their primary energy sources, to form the core of classical energy transduction or bioenergetics.

Energy-transducing membranes possess a number of distinguishing features. Each membrane has two distinct types of proton pump. The nature of the primary proton pump depends on the energy source used by the membrane; in the case of mitochondria or respiring bacteria an electron-transfer chain catalyses the 'downhill' transfer of electrons from substrates to final acceptors such as O_2 and uses this energy to generate a gradient of protons (details of this will be covered in Chapter 5). Photosynthetic bacteria exploit the energy available from the absorption of quanta of visible light to generate a gradient of protons, while chloroplasts not only accomplish this but also drive electrons 'uphill' from water to acceptors such as $NADP^+$ (Chapter 6). It is a useful convention to define the side of the membrane to which protons are pumped as the P or positive side and the side from which they have originated as the N or negative side (Fig. 1.1).

In contrast to the variety of primary proton pumps, all energy-conserving membranes contain a highly conserved secondary proton pump. If this pump, termed the ATP synthase or the proton-translocating ATPase (Chapter 7), were operating in isolation in a membrane it would hydrolyse ATP to ADP and Pi and pump protons in the same direction as the primary pump (Fig. 1.1). However, the essence of the chemiosmotic theory is that the primary proton pump generates such a high gradient of protons that it forces the secondary

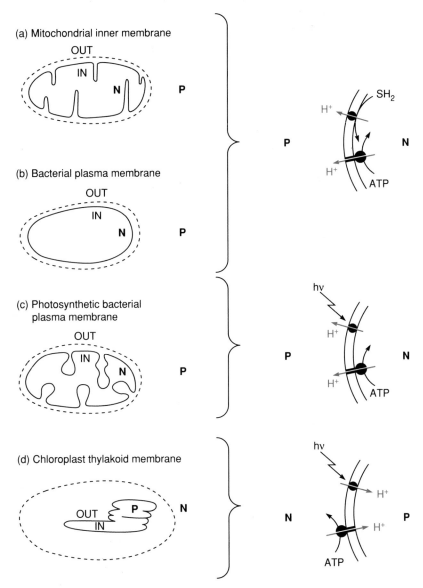

Figure 1.1 Energy-transducing membranes contain pairs of proton pumps with the same orientation.

In each case the primary pump (utilizing either electrons or photons $h\nu$) pumps protons from the N (negative) compartment to the P (positive) compartment. Note that the ATP synthase in each case is shown acting in the direction of ATP hydrolysis, when it would also pump protons from the N- to the P-phase.

pump to reverse and actually synthesize ATP from ADP and Pi. It should be noted for both the primary and secondary pumps that metabolism (i.e. electron flow or phosphorylation) is tightly coupled to proton translocation: the one cannot occur without the other.

What is the nature of this gradient of protons? The quantitative thermodynamic measure is the proton electrochemical gradient $\Delta\tilde{\mu}_{H+}$. An ion electrochemical gradient, expressed in kJ mol^{-1}, is a thermodynamic measure of the extent to which an ion gradient is removed from equilibrium (and hence capable of doing work) and will be derived in Chapter 3. For the present it is sufficient to note that $\Delta\tilde{\mu}_{H+}$ has two components: one due to the concentration difference of protons across the membrane, ΔpH, and one due to the difference in electrical potential between the two aqueous phases separated by the membrane, the membrane potential, $\Delta\psi$. A bioenergetic convention is to convert $\Delta\tilde{\mu}_{H+}$ into units of electrical potential, i.e. millivolts, and to refer to this as the *protonmotive force*, expressed by the symbol Δp.

In only a few cases, such as the chloroplast, does Δp exist solely as a pH difference across the energy-conserving membrane. In this example the pH

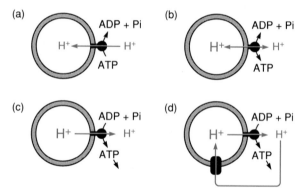

Figure 1.2 A hypothetical model to demonstrate chemiosmotic coupling. An ATP synthase complex is incorporated into a phospholipid membrane such that the ATP binding site is on the outside. (a) ATP is added, the nucleotide starts to be hydrolysed to ADP + Pi and protons are pumped into the vesicle lumen. As ATP is converted to ADP + Pi the energy available from the hydrolysis steadily decreases, while the energy required to pump further protons against the gradient which has already been established steadily increases. (b) Eventually an equilibrium is attained. (c) If this equilibrium is now disturbed, for example by removing ATP, the ATP synthase will reverse and attempt to re-establish the equilibrium by synthesizing ATP. Net synthesis, however, would be very small as the gradient of protons would rapidly collapse and a new equilibrium would be established. (d) For continuous ATP synthesis a primary proton pump, driven in this example by photons ($h\nu$), is required to pump protons across the same membrane and replenish the gradient of protons. A proton circuit has now been established. This is what occurs across energy-conserving membranes: ATP is continuously removed for cytoplasmic ATP-consuming reactions, while the gradient of protons, Δp, is continuously replenished by the respiratory or photosynthetic electron-transfer chains.

gradient, ΔpH, across the thylakoid membrane can exceed 3 units. Although the thylakoid space is therefore highly acidic, there are no enzymes in this compartment which might be compromised by the low pH. The more common situation is where $\Delta\psi$ is the dominant component and the pH gradient is small: perhaps only 0.5 pH units. This occurs, for example, in the mitochondrion, allowing enzymes in both the mitochondrial matrix and cell cytoplasm to operate close to neutral pH.

Figure 1.2 constructs a hypothetical ATP-synthesizing organelle from first principles. A central feature is the *proton circuit* linking the primary pump with the ATP synthase. This proton circuit (Fig. 1.3) is closely analogous to an electrical circuit, and the analogy holds even when discussing detailed and complex energy flows (see Chapter 4). As with the electrical case one can measure a potential, the protonmotive force, Δp, a current of protons, J_{H+}, and a conductance for protons, C_MH^+, defined from the current flowing through a component divided by the potential drop across the component. These aspects of the proton circuit will be discussed in Chapter 4.

To avoid short-circuiting, it is evident that the membrane must be closed and must possess a high resistance to protons. Protonophores, also called *uncouplers*, are synthetic compounds which break the energetic coupling between the ATP synthase and the primary pump. Uncouplers were described long before the chemiosmotic theory was propounded, and one of the most

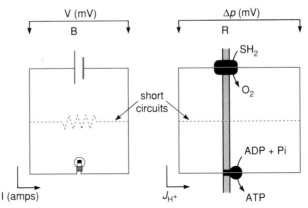

Figure 1.3 Proton circuits and electrical circuits are analogous in the following respects.
(a) Both have generators of potential difference (the battery, B, and the respiratory chain, R, respectively). (b) Both potentials (voltage difference and Δp) can be expressed in mV. (c) Both potentials can be used to perform useful work (the light bulb and ATP synthesis respectively). (d) Both circuits can be short-circuited (by respectively a piece of wire or a protonophore – an agent which makes membranes permeable to protons, Section 2.3.5a). (e) The rate of chemical conversion in the battery (or respiratory chain) is tightly linked to the current of electrons (or protons) flowing in the rest of the circuit, which in turn depends on the resistance of the circuit. (f) The potentials fall as the currents drawn increase (J denotes proton flux).

successful predictions was that uncouplers act by increasing the proton conductance of the membrane and inducing just such a short-circuit (Fig. 1.3).

Mitochondrial and bacterial membranes have not only to maintain a proton circuit across their membranes, but must also provide mechanisms for the uptake and excretion of ions and metabolites. Since it is energetically unfavourable for a negatively charged metabolite to enter the negative interior of a mitochondrion or bacterium, the chemiosmotic theory also included a proposal for the existence of transport systems in which metabolites were transported together with protons, or by an equivalent exchange with OH^-. Alternatively, components of Δp could be exploited in other ways so as to drive transport (see Chapter 8).

There is some semantic confusion as to what constitutes the 'chemiosmotic theory' or the 'Mitchell theory'. In this book the term 'chemiosmosis' will be used synonymously with the central concept outlined above, namely that coupling occurs through the intermediacy of a proton electrochemical gradient (protonmotive force, Δp), or, in the case of some bacteria, an equivalent Na^+ circuit. There are many secondary hypotheses, from Peter Mitchell and others, as to the precise molecular mechanism by which the pumps operate, but it must be emphasized that the validity of the theory is independent of the fate of these secondary theories.

1.3 THE MORPHOLOGY OF ENERGY-TRANSDUCING ORGANELLES

1.3.1 Mitochondria and submitochondrial particles

The appearance of a section through a typical mitochondrion is shown in Fig. 1.4 . Mitochondria are typically 0.7–1 μm in length. Their shape is not fixed but varies continuously in the cell, and the appearance of the cristae can be quite different in mitochondria isolated from different tissues or even with the same mitochondria suspended in different media. Thus heart mitochondria, for which periods of high respiratory activity are required, tend to have a greater surface area of cristae than liver mitochondria. In the past, ultrastructural changes observed for isolated mitochondria under different conditions have been thought to be directly related to the mechanism of energy transduction. This is no longer thought to be the case. There is evidence for larger and filamentous mitochondria in some cell types. This aspect and the possible implications for energy transmission within the cell are discussed by Skulachev (1988) and will not be treated in this book.

The outer mitochondrial membrane possesses proteins termed porins which act as non-specific pores for solutes of molecular mass less than 10 kDa, and is therefore freely permeable to ions and most metabolites. The inner membrane is energy transducing. In mitochondrial preparations which have been negatively stained with phosphotungstate, knobs, the part of the ATP synthase where adenine nucleotides and phosphate bind, are visible on the matrix

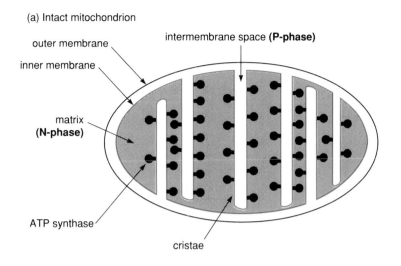

(a) Intact mitochondrion

intermembrane space **(P-phase)**

outer membrane

inner membrane

matrix
(N-phase)

ATP synthase

cristae

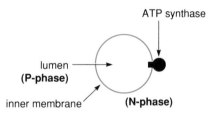

(b) Sub-mitochondrial particle (SMP)

ATP synthase

lumen
(P-phase)

inner membrane   **(N-phase)**

Figure 1.4   Mitochondria and submitochondrial particles (SMPs).
P and N refer to the positive and negative compartments. This nomenclature has the advantage that the ATP synthase is at the N-side of the
membrane in both mitochondria and sub-mitochondrial particles.

face (N-side – Fig. 1.4) of the inner membrane. The enzymes of the citric acid
cycle are in the matrix, except for succinate dehydrogenase, which is bound to
the N-face of the inner membrane. It must be borne in mind that the concentration of protein in the matrix can approach 500 mg/ml and there may be
a considerable structural organization within this enormously concentrated
solution. The matrix pools of NAD^+ and $NADP^+$ are separate from those in
the cytosol, while matrix ADP and ATP communicate with the cytoplasm
through the adenine nucleotide exchanger. Additionally, specific carrier proteins exist for the transport of many metabolites.

Mitochondria are usually prepared by gentle homogenization of the tissue
in isotonic sucrose, followed by differential centrifugation to separate
mitochondria from nuclei, cell debris and microsomes (fragmented
endoplasmic reticulum). Although this method is effective with fragile tissues
such as liver, tougher tissues such as heart must either first be incubated with
a protease, such as nagarse, or be exposed briefly to a blender to break the
muscle fibres. Yeast mitochondria are isolated following digestion of the cell
wall with snail-gut enzyme.

Ultrasonic disintegration of mitochondria produces inverted submitochondrial particles (SMPs) (Fig. 1.4). Because these have the substrate binding sites for both the respiratory chain and the ATP synthase on the outside, they have been much exploited for investigations into the mechanism of energy transduction.

1.3.2 Respiratory bacteria and derived preparations

Energy transduction in bacteria is associated with the cytoplasmic membrane (Fig. 1.5). In Gram-negative bacteria (which are typically of similar size to mitochondria) this membrane is separated from a peptidoglycan layer and an outer membrane by the periplasm, which is approximately 10 nm wide. In Gram-positive bacteria the periplasm is absent and the cell wall is closely juxtaposed to the cytoplasmic membrane. Fig. 1.5 is an oversimplification because in some organisms with a very high rate of respiration there are substantial infoldings of the cytoplasmic membrane.

It is difficult to study energy transduction with intact bacteria because:

● Many reagents do not penetrate the outer membrane of Gram-negative organisms.

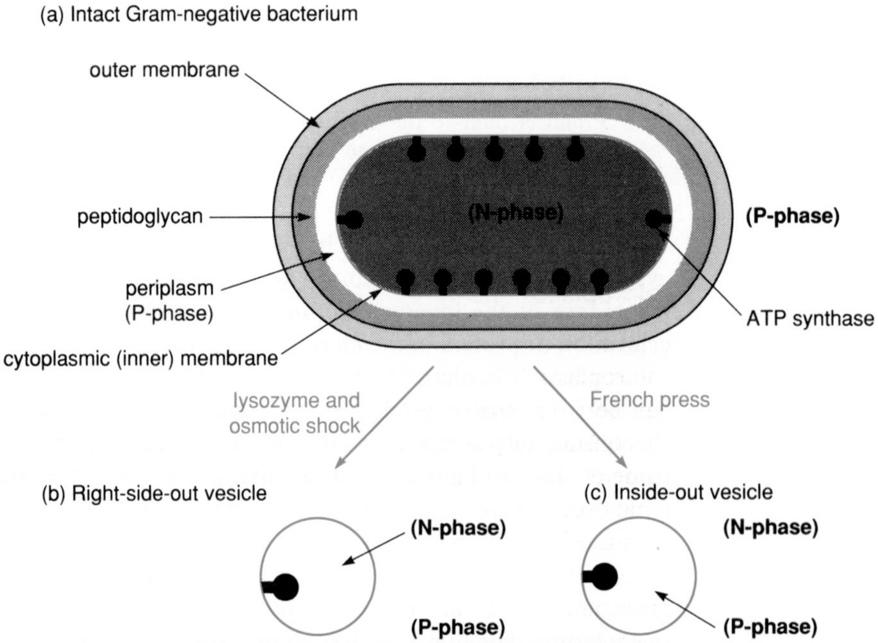

Figure 1.5 Gram-negative bacteria and derived preparations.
P and N refer to positive and negative compartments. The periplasm is part of the P-phase. Note that Gram-positive bacteria differ by lacking an outer membrane and a periplasm. Nevertheless, similar vesicle preparations can be made from these organisms.

- ADP, ATP, NAD$^+$ and NADH do not cross the cytoplasmic membrane.
- Cells are frequently difficult to starve of endogenous substrates and thus there can be ambiguity as to the substrate which is donating electrons to a respiratory chain.
- Study of transport can be complicated by subsequent metabolism of the substrate.

Cell-free vesicular systems can overcome these problems. For most transport studies *right-side-out vesicles* are required. These can often be obtained by weakening the cell wall, e.g. with lysozyme, and then exposing the resulting spheroplasts or protoplasts to osmotic shock. Vesicles with this orientation can only oxidize substrates that have an external binding site or can permeate the cytoplasmic membrane. They cannot hydrolyse or synthesize ATP, in contrast to *inside-out vesicles* which can frequently be prepared by extruding cells at very high pressure through an orifice in a French press. These vesicles can oxidize NADH and phosphorylate added ADP. The method of vesicle preparation varies between genera; occasionally osmotic shock may give inside-out vesicles or a mixture of the two orientations. This last feature need not be a major problem because, for example, in a study of ATP synthesis the reaction would be confined to the inside-out population. Nevertheless, failure to characterize the orientation of vesicles has caused confusion in the past.

Vesicle preparations have some disadvantages, such as the loss of periplasmic electron-transport or solute-binding proteins; the latter play key roles in many aspects of bacterial energy transduction (Ferguson, 1992) (Chapters 5 and 8). Also the membrane of a vesicle may be somewhat leaky with the result that the stoichiometry of an energy-transduction reaction may be adversely affected.

1.3.3 Chloroplasts and their thylakoids

Chloroplasts are plastids, organelles peculiar to plants (Fig. 1.6) ; there may be from one to a hundred or more chloroplasts per cell. Chloroplasts are considerably larger than the average mitochondrion, being 4–10 μm in diameter and 1–2 μm thick and bounded by an envelope of two closely juxtaposed membranes, the matrix within the inner membrane being the stroma (Fig. 1.6). Within stroma are flattened vesicles called thylakoids, the membranes of which have regions that are folded so that the contiguous membrane has a stacked appearance, referred to as the grana (Fig. 1.6). Energy conservation occurs across the thylakoid membranes and light causes the translocation of protons into the internal thylakoid spaces (usually called the lumen). The chloroplast ATP synthase is part of the thylakoid membrane and is orientated with its 'knobs' on the stromal face of the membrane. Thus the lumen space inside the thylakoid is the P-compartment and the stroma the

(a) Intact chloroplast

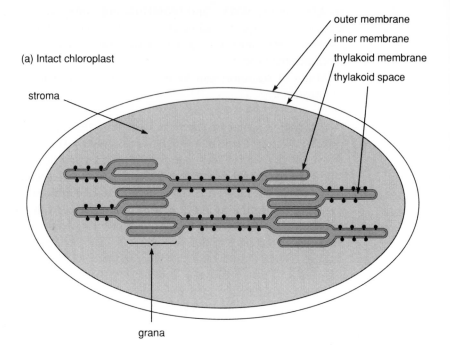

outer membrane
inner membrane
thylakoid membrane
thylakoid space

stroma

grana

(b) Detail of the thylakoid

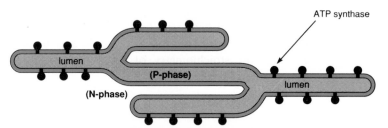

ATP synthase

lumen
(P-phase)
(N-phase)
lumen

Figure 1.6   Chloroplasts and their thylakoids.
(a) 'Intact' chloroplast; note that it is probable that there is a single continuous thylakoid space. (b) Detail of the thylakoid. The membrane is heterogeneous with, for example, the ATP synthase being excluded from the grana (appressed regions) where the membrane is closely stacked. Light-driven proton pumping occurs from the N- to the P-phase (note, however, that in steady-state light the membrane potential across a thylakoid membrane is negligible).

N-compartment. The ATP and NADPH generated by photosynthetic phosphorylation are used by the CO_2-fixing dark reactions of the Calvin cycle located in the stroma.

Although at first sight the structure of chloroplasts appears to be very different from that of mitochondria, the only topological distinction is that the thylakoids, in contrast to the mitochondria cristae, can be thought of as having become separated from the inner membrane, with the result that the thylakoid lumen is a separate compartment, unlike the 'cristal space', which is continuous with the intermembrane space of mitochondria.

Chloroplasts are prepared by gentle homogenization of leaves (e.g. from peas, spinach or lettuce), but avoiding material rich in polyphenols or acid, in isotonic sucrose or sorbitol. After removal of cell debris, the chloroplasts are sedimented by low-speed centrifugation. A rapid and careful preparation will contain a high proportion of intact chloroplasts capable of high rates of CO_2 fixation. Slightly harsher conditions yield 'broken chloroplasts' which have lost the envelope membranes and the stroma contents. These broken chloroplasts (thylakoid membrane preparations) do not fix CO_2 but are capable of high rates of reduction of added electron acceptors and of photophosphorylation. They are often the choice material for bioenergetic investigations because the chloroplast envelope prevents access of substances such as ADP or $NADP^+$.

1.3.4 Photosynthetic bacteria and chromatophores

Three groups of prokaryotes catalyse photosynthetic electron transfer: the green bacteria, the purple bacteria, and the cyanobacteria (or blue-green algae). The purple bacteria are divided into two groups: the Rhodospirillaceae (or non-sulphur) and the Chromatiaceae (or sulphur). Cyanobacteria carry out non-cyclic electron transfer (Chapter 6), and use H_2O as electron donor, and are in this respect similar to chloroplasts. Of the remaining groups, the purple bacteria, and especially the Rhodospirillaceae, have been the more intensively investigated, and several factors make them suitable for bioenergetic studies. First, mechanical disruption of the cells (for example in a French press) enables the characteristic invaginations of the cytoplasmic membrane to bud off and form isolated closed vesicles called chromatophores (Fig. 1.7).

Chromatophores retain the capacity for photosynthetic energy transduction and possess the same orientation as the inside-out vesicles discussed in Section 1.3.2. Light-driven ATP synthesis can be studied, and they have been important for chemiosmotic studies, especially since they are so small (diameters of the order of 50 nm) that light scattering is negligible and suspensions are optically clear. A further advantage of the purple bacteria is that the reaction centres (the primary photochemical complexes) can be readily isolated (Chapter 6). Finally these organisms will grow in the dark, for example by aerobic respiration, permitting the study of mutants defective in the photosynthetic apparatus.

In addition to these bacteria, halobacteria carry out a unique light-dependent energy transduction in which a single protein, bacteriorhodopsin, acts as a light-driven proton pump (Chapter 6).

1.3.5 Reconstituted systems

An essential feature of the chemiosmotic theory is that the primary and secondary proton pumps should be functionally and structurally separable. In

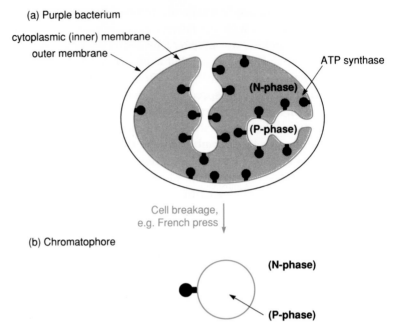

Figure 1.7   Preparations derived from photosynthetic bacteria.
(a) Purple bacterium. (b) Chromatophore. Note that these bacteria are
Gram-negative and possess peptidoglycan and periplasm (cf. Fig. 1.5)
which are here omitted for clarity.

order to observe proton translocation, purification of proton-translocating
complexes must be followed by their reincorporation into synthetic, closed
membranes that have low permeabilities to ions. The value of such 'reconsti-
tutions' is two-fold. First, it allows aspects of the chemiosmotic theory to be
tested (e.g. Do all energy-transducing complexes pump protons? Does each
complex pump protons as an autonomous unit?) Second, it allows the
minimum functional unit to be established as a preliminary to investigating the
mechanism, for example, of a pump or a protein that catalyses active
transport.

Membrane proteins such as those involved in proton translocation
(pumping) are generally purified following solubilization of the membrane
with a suitable detergent which is usually non-ionic. Once purified there are
two principal ways in which they can be reconstituted into a membrane structure. The first is to mix the purified protein dispersed in a suitable detergent,
preferably one with a high critical micelle concentration, with phospholipids
and then to allow the concentration of detergent to fall slowly either by dialysis
or gel filtration. Under optimal conditions this can lead to the formation of
unilamellar phospholipid vesicles. The protein can in principle be oriented in
either of two ways (Fig 1.8). If the protein uses a substrate, e.g. ATP, to which
the phospholipid bilayer is impermeable, then mixed orientation is not a
problem because only those molecules with their catalytic site facing outward

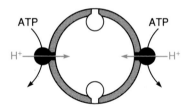

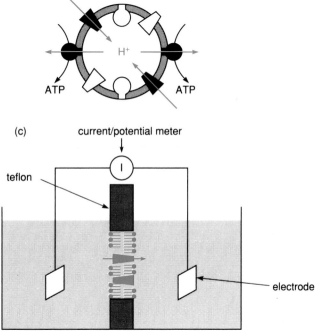

Figure 1.8 Principal systems for the reconstitution of proton-translocating complexes.
(a) Unilamellar phospholipid vesicles containing, as an example, the ATP synthase enzyme. Although the protein is shown as reconstituting with random orientation with respect to the two sides of the membrane, only the enzymes with the active site on the outer surface are accessible to added substate. (b) Unilamellar phospholipid vesicles containing two proton pumps. A functional proton circuit can form if the two pumps not only reconstitute into the same vesicle but also have the appropriate respective orientations. (c) The black-lipid membrane system in which a protein is incorporated into a bilayer formed over a small aperture that separates two macroscopic compartments into which electrodes can be inserted. Asymmetric operation of a protein that has incorporated randomly can be achieved by adding substrate to just one of the compartments.

will be accessible to the substrate. On the other hand, if the protein is a photosynthetic system then asymmetry can obviously not be imposed in this way.

Fortunately, proteins frequently orient asymmetrically; the differences in radius of curvature for the two sides of the vesicle may be an important factor. A more demanding type of reconstitution is when the presence of two different proteins (e.g. a primary and secondary pump) is required in the same membrane. The problem here is to ensure not only that at least a majority of vesicles contain both proteins, but also that the relative orientations of the two proteins allow coupling between them via the proton circuit (Fig. 1.8 and cf. Figs 1.2 and 1.3).

A second procedure for reconstitution is to incorporate the purified protein into a planar bilayer which can be formed over a tiny orifice that separates two reaction chambers (Fig. 1.8c). The insertion of protein is frequently achieved by fusing phospholipid vesicles containing the protein of interest with the planar bilayer. Alternatively, in some cases it has been possible to form the bilayer directly by application of a protein phospholipid mixture in a suitable volatile solvent to the aperture. The amount of enzyme incorporated into such bilayers is usually so small that biochemical or chemical assays of activity are not possible. However, the crucial advantage of this type of system is that macroscopic electrodes can be inserted into the two chambers and thus direct electric measurements (either current or voltage) of any ion or electron movements driven by the reconstituted protein can be made.

1.4 A BRIEF HISTORICAL BACKGROUND

Before the role of the proton electrochemical gradient (or protonmotive force, Δp) was understood, the search for the 'energy-transducing intermediate' linking pairs of protein assemblies (i.e. primary and secondary proton pumps, see Section 1.2) provided the central stimulus for bioenergetic research in the 1950s and 1960s. The intermediate proved to be highly elusive, and by the mid-1960s uncertainty was so extreme that it prompted the statement that 'anybody who is not thoroughly confused just doesn't understand the problem'. However, despite the central enigma, by that time the main energy-transducing pathways within the mitochondrion had been established. Any theory had to be consistent with a number of basic observations.

(a) The electron-transfer chain comprises a sequence of electron carriers with three separate 'regions' where redox energy can be conserved in the synthesis of ATP.
(b) The rate of respiration is controlled by the demand for ATP (respiratory control).
(c) Coupling between respiration and ATP synthesis can be disrupted by a group of agents termed 'uncouplers' which abolish respiratory control (i.e. stimulate respiration in the absence of ATP synthesis) and allow

mitochondria to catalyse a rapid ATP hydrolysis ('uncoupler-stimulated ATPase').

(d) The antibiotic oligomycin (Section 7.2) inhibits both the synthesis and uncoupler-stimulated hydrolysis of ATP.

(e) The energy from respiration can be coupled not only to the synthesis of ATP but also to the accumulation of Ca^{2+} and to the reduction of NAD^+ and $NADP^+$.

(f) The processes specified in (e) can be driven by the hydrolysis of ATP in anaerobic mitochondria, when they can be inhibited by both uncouplers and oligomycin.

These observations (and related ones for thylakoids and bacteria) were consistent with pathways of energy transduction radiating from a common 'energy pool'. Figure 1.9 shows a current model. By analogy to the only reactions in which a detailed mechanism of ATP synthesis was available, namely the phosphorylations (traditionally known as substrate-level phosphorylation) in glycolysis and the citric acid cycle in which thio-esters participate as reaction intermediates, it was reasonable to expect that the common energy-transducing entity would be a chemical 'high-energy' intermediate, usually referred to by the short-hand 'squiggle'. Despite a number of false starts, no 'squiggle' was ever found.

As so often in science, the solution was simple and elegant. Peter Mitchell,

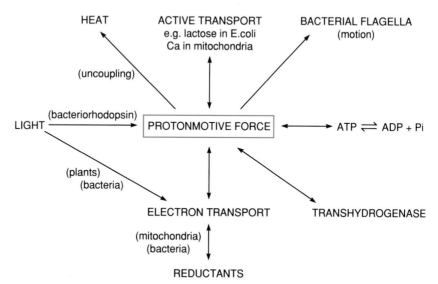

Figure 1.9 Pathways of energy transduction.
The common 'energy pool' was originally thought to be a chemical compound 'squiggle' formed during electron transfer through the respiratory chain, but was proposed by the chemiosmotic theory to be the protonmotive force, Δp. The wide range of different coupled activities now recognized would be very difficult to reconcile with a localized chemical species as originally envisaged for 'squiggle'.

whose background was in transport rather than classical bioenergetics, made what seemed at the time an extraordinary suggestion that the only 'intermediate' was a proton electrochemical gradient across the membrane. The early theoretical papers by Mitchell were not readily comprehended, and the subsequent, frequently acrimonious, debate between advocates of conflicting hypotheses continued unabated for 15 years and has even been the subject of a sociological study (Gilbert and Mulkay, 1984). As a result the chemiosmotic 'hypothesis' has probably been submitted to a more rigorous trial than any other comparable biochemical theory. The steady flow of converts to the hypothesis was ultimately recognized with the award to Peter Mitchell of the Nobel Prize for Chemistry in 1978.

1.5 MITCHELL'S POSTULATES

Mitchell proposed four basic requirements for the experimental verification of the concept of proton electrochemical coupling, and some of the more important experiments will now be reviewed briefly. Further details and additional evidence in support of Mitchell's postulates follow in the subsequent chapters.

1.5.1 The respiratory and photosynthetic electron-transfer chains should translocate protons

Energy-dependent proton translocation was first reported in 'broken' chloroplast preparations (Neumann and Jagendorf, 1964). Mitochondrial proton translocation was shown by Mitchell and Moyle (1965, 1967a) by adding a small pulse of O_2 to an anaerobic suspension and following the acidification of the medium. Low concentrations of uncouplers accelerated the decay of the acidification phase following anoxia. A stoichiometry of proton extrusion of about $2H^+/2e^-$ per proton-translocating respiratory chain complex was found (but see Chapter 4 for the contemporary view on the stoichiometry). During continuous respiration a Δp of over 200 mV was found (Mitchell and Moyle, 1969) (Section 4.2). The direction of proton translocation was reversed in inverted SMPs (Mitchell and Moyle, 1965). Skulachev (1970) showed that mitochondria could use the membrane potential generated by proton pumping to accumulate a wide variety of synthetic lipophilic cations, while SMPs and chromatophores, both with inverted polarity, accumulated the corresponding anions. Finally, demonstrations that purified respiratory complexes could function as autonomous proton pumps in reconstituted systems (Section 5.2) proved inconsistent with the alternative 'chemical' explanation of a 'squiggle'-driven common proton pump.

1.5.2 The ATP synthase should function as a reversible proton-translocating ATPase

Injection of a small amount of ATP into a suspension of anaerobic mitochondria resulted in an expulsion of protons followed by a slow decay which could be accelerated by uncouplers (Mitchell and Moyle, 1967a). Purified reconstituted ATP synthase catalysed a similar proton translocation (Kagawa *et al.* 1973). ATP hydrolysis could maintain a Δp in excess of 200 mV, e.g. Nicholls (1974a), while a pH gradient imposed in the dark could drive ATP synthesis by thylakoids (Jagendorf and Uribe, 1966).

1.5.3 Energy-transducing membranes should have a low effective proton conductance

A low proton permeability can be inferred from the parallel actions of agents which induce proton permeability in synthetic bilayers and at the same time uncouple mitochondria. Proton permeability was measured more directly by Mitchell and Moyle (1967b) by following the rate at which a pH gradient induced by addition of HCl to the medium decayed as protons entered the mitochondrial matrix. Brown fat mitochondria (Section 4.7) were found to control their heat production by modulating their proton conductance (Nicholls, 1974b).

1.5.4 Energy-transducing membranes should possess specific exchange carriers to permit metabolites to permeate, and osmotic stability to be maintained, in the presence of a high membrane potential

The decay of ΔpH following a burst of respiration is accelerated by anions such as succinate, Pi and malonate, and by Na^+ (Mitchell and Moyle, 1967a), suggesting that anions may be transported with protons and Na^+ in exchange for protons (see Fig. 4.8). From the rate of swelling of non-respiring mitochondria in ammonium salts, Chappell and colleagues (Chappell, 1968) obtained evidence for a number of transport systems in which anions are largely transported as uncharged species, either together with protons, or in exchange for other anions (Section 8.4).

1.6 CONCLUDING REMARKS

The aim of this opening chapter is to establish the basic material with which this book is concerned. Figure 1.9 presents in summary form the wide range of processes that are coupled by a protonmotive force in bacteria, mitochondria and thylakoids. Some significant principles and points of experimental design have been put to one side in order to present such an overview. This important material is dealt with in the remaining chapters, which are also concerned with contemporary information about the molecular nature of the pumps and their mechanisms.

The malate–aspartate shuttle enables reducing equivalents from cytosolic NADH to be transported into the mitochondrial matrix (Fig. 8.6)

2 ION TRANSPORT ACROSS ENERGY-CONSERVING MEMBRANES

2.1 INTRODUCTION

The chemiosmotic theory requires that the transport of ions be considered an integral part of bioenergetics. It was this more than any other single factor which helped to remove the artificial distinction, present in the early days of bioenergetics, between events occurring in energy-conserving membranes, which were considered the valid preserve of the bioenergeticist, and closely related transport events occurring in the eukaryotic plasma membrane, endoplasmic reticulum, secretory vesicles, etc., which were assigned to a different field of research.

In this chapter we shall describe the basic permeability properties of membranes and the abilities of ionophores to induce additional pathways of ion permeation, bearing in mind that what is discussed is equally applicable to energy-conserving and non-energy-conserving membranes – after all, the only significant distinction between the two is the presence of the proton pumps in the former.

For an ion to be transported across a membrane both a driving force and a pathway are required. Driving forces can be metabolic energy (such as ATP hydrolysis), concentration gradients, electrical potentials, or combinations of these. These forces will be discussed in Chapter 3; this chapter will deal with the natural and induced pathways which occur in energy-conserving membranes.

2.2 THE CATEGORIZATION OF ION TRANSPORT

In order to reduce the complexity of membrane transport events, it is useful to categorize any transport process in terms of a number of simple criteria. Any transport process can be described in terms of the following four criteria:

1. Does transport occur across the bilayer or is it protein-catalysed (Fig. 2.1a)?
2. Is transport passive or *directly* coupled to metabolism (Fig. 2.1b)?
3. Does a transport process involve a single ion or metabolite, or are fluxes *directly* coupled together (Fig. 2.1c)?
4. Does the transport process involve charge transfer across the membrane (Fig. 2.1d)?

2.2.1 Bilayer-mediated versus protein-catalysed transport

A consequence of the fluid-mosaic model of membrane structure is that transport can from first principles occur either through lipid bilayer regions of the membrane or be catalysed by integral, membrane-spanning proteins. The distinction between protein-catalysed transport and transport across the bilayer regions of the membrane is fundamental and will be emphasized in this chapter.

While the fluid-mosaic model is usually represented with protein 'icebergs' floating in a sea of lipid, the high proportion of protein in energy-conserving membranes (in the case of the mitochondrial inner membrane 50% of the membrane is integral protein, 25% peripheral protein and 25% lipid) results in a relatively close packing of the proteins. Unlike with plasma membrane proteins, no attachment to cytoskeletal elements occurs.

Consistent with the proposal that mitochondria and chloroplasts evolved from respiring or photosynthetic bacteria, energy-conserving membranes tend to have distinctive lipid compositions: 10% of the mitochondrial inner membrane lipid is cardiolipin, while only 16% of the chloroplast thylakoid membrane lipid is phospholipid, the remainder being galactolipids (40%) sulpholipids (4%) and photosynthetic pigments (40%).

Despite this heterogeneity of lipid composition, the native and ionophore-induced permeability properties of the bilayer regions of the different membranes are sufficiently similar to justify extrapolations to energy-transducing membranes from artificial bilayer preparations. However, protein-catalysed transport can be unique, not only to a given organelle but also to an individual tissue, depending on the genes expressed in that cell. For example the inner membrane of rat liver mitochondria possesses protein-catalysed transport properties which are absent in mitochondria from rat heart (Section 8.4).

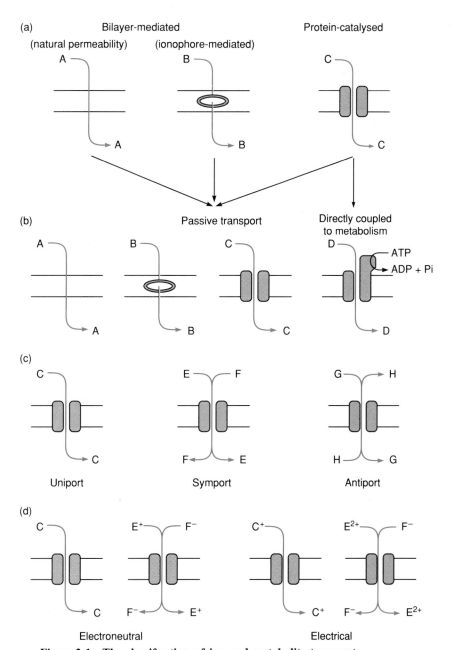

Figure 2.1 The classification of ion and metabolite transport.
(a) Transport may be bilayer-mediated (via either natural permeation across the membrane (A) or an ionophore-induced pathway (B)) or protein-catalysed (C). (b) Transport by any of the three pathways in (a) can be passive (not directly coupled to metabolism (A, B, C)) or, in the case of protein-catalysed transport alone, directly coupled to metabolism, e.g. ATP hydrolysis (D). (c) Transport, by any of the pathways in (b), may occur of a single species (C), of two or more ions whose transport is tightly coupled together by symport (or co-transport), (E:F) or by antiport (or exchange diffusion), (G/H). (d) Any of the stoichiometries in (c) may be electroneutral (C, $E^+:F^-$) or electrical (electrogenic, electrophoretic), (C^+, E^{2+}/F^-).

147,002

LIBRARY
College of St. Francis
JOLIET. ILLINOIS

2.2.2 Transport directly coupled to metabolism versus passive transport

A tight coupling of transport to metabolism occurs in the ion pumps which are central to chemiosmotic energy transduction. The term 'active transport' is often applied in this context, although it is important to restrict the term to examples where a direct coupling occurs (e.g. Fig. 2.1b), and *not* to include all cases where ions are concentrated across a membrane. Thus ions can also be accumulated without direct metabolic coupling if there is a membrane potential or if transport is coupled to the 'downhill' movement of a second ion. For example, while Ca^{2+} is accumulated within the sarcoplasmic reticulum by an ion pump (the Ca^{2+}-ATPase), the same ion is accumulated across the mitochondrial inner membrane as a consequence of the membrane potential (Section 8.3). Only the former could be strictly described as 'active'; mitochondrial Ca^{2+} accumulation occurrs down the electrochemical gradient for this ion. In some texts the terms primary and secondary active transport are used to distinguish these examples. However, it is the case that the term 'active transport' is often used to categorize any process in which a concentration gradient of a solute or ion is established across a membrane. Naturally, only protein-catalysed transport can be directly coupled to metabolism (Fig. 2.1b).

2.2.3 Uniport, symport and antiport

The molecular mechanism of a transport process can involve a single ion or the coupled transport of two or more species (Fig. 2.1c). A transport process involving a single ion is termed a uniport. Examples of uniports include the uptake pathway for Ca^{2+} across the inner mitochondrial membrane (Section 8.3) and the proton permeability induced in bilayers by the addition of proton translocators such as dinitrophenol (Section 2.3.5a). A transport process involving the obligatory coupling of two or more ions in parallel is termed symport or co-transport. In this book we shall use the shorthand A:B to denote symport of A and B. A number of cases of H^+:metabolite symport occur across the bacterial membrane (Section 8.5.1). The equivalent tightly coupled process where the transport of one ion is linked to the transport of another species in the opposite direction is termed antiport or exchange-diffusion (Fig. 2.1c). Examples include the Na^+/H^+ antiport activity which is present in the inner mitochondrial membrane (Section 8.2) and the K^+/H^+ antiport catalysed by the ionophore nigericin in bilayers (Section 2.3.4a). If one of the ions involved in a nominal symport or antiport mechanism is a H^+ or OH^-, it is usually impossible to distinguish between the symport of a species with a H^+ and the antiport of the species with an OH^-. For example, the mitochondrial phosphate carrier (Section 8.4.4) may be variously represented as a Pi^-/OH^- antiport or a H^+:Pi^- symport.

Closely related transport pathways exist across non-energy-conserving membranes. At the plasma membrane the Na^+ ion can be involved in uniport (through a voltage-activated channel), symport (e.g. Na^+:glucose co-transport), and antiport (e.g. the $3Na^+/Ca^{2+}$ exchanger), while more complex stoichiometries may occur — for example some neuronal membranes possess a carrier which catalyses the co-transport of Na^+ and glutamate coupled to the antiport of a third ion, K^+ (Nicholls and Attwell, 1990).

2.2.4 Electroneutral versus electrical transport

Electroneutral transport involves no net charge transfer across the membrane. Transport may be electroneutral either because an uncharged species is transported by a uniport, as the result of the symport of a cation and an anion or the antiport of two ions of equal charge (Fig. 2.1d), an example of the last being the K^+/H^+ antiport catalysed by nigericin. Electrical transport is frequently termed either electrogenic ('creating a potential'; proton pumping driven by ATP hydrolysis would be an example) or electrophoretic ('moving in response to a pre-existing potential'; Ca^{2+} uniport into mitochondria would be an example). As these terms can refer to the same pathway observed under different conditions the overall term 'electrical' will be used here.

It is important to distinguish between movement of charge at the molecular level, as discussed here, and the overall electroneutrality of the total ion movements across a given membrane. The latter follows from the impossibility of separating more than minute quantities of positive and negative charge across a membrane without building up a large membrane potential. Thus the separation of 1 nmol of charge across the inner membranes of 1 mg of mitochondria results in the build-up of more than 200 mV of potential. Or put another way a single turnover of all the electron-transport components in an individual mitochondrion or bacterium will translocate sufficient charge to establish a membrane potential approaching 200 mV. The establishment of such potentials by the movement of so little charge is a consequence of the low electrical capacitance of biological membranes (typically estimated as $1\ \mu F\ cm^{-2}$). However, this property does not preclude the occurrence of steady state electrical events at the molecular level as long as these compensate each other (Section 2.5). In addition it is necessary to appreciate that the effect on an energy-transducing membrane of a tightly coupled electroneutral antiporter is not the same as that caused by the addition of two electrical uniporters for the same ions.

The four criteria discussed above allow a comprehensive description of a transport process; for example, proton pumping by the ATP synthase is an example of a protein-catalysed, metabolism-coupled, electrical uniport, while the ionophore nigericin (Section 2.3.4a) catalyses a bilayer-mediated, passive, electroneutral antiport.

2.3 BILAYER-MEDIATED TRANSPORT

2.3.1 The natural permeability properties of bilayers

The hydrophobic core possessed by lipid bilayers creates an effective barrier to the passage of charged species. With a few important exceptions (Section 2.3.6) cations and anions do not permeate bilayers. This impermeability extends to the proton, and this property is vital for energy transduction to avoid short-circuiting the proton circuit. Not only does the bilayer have a high electrical resistance, but it can also withstand very high electrical fields. An energy-conserving membrane with a membrane potential of 200 mV across it has an electrical field in excess of 300 000 V cm^{-1} across its hydrophobic core.

A variety of uncharged species can cross bilayers. H_2O, O_2 and CO_2 are all highly permeable, as are the uncharged forms of a number of low molecular weight acids and bases, such as ammonia and acetic (ethanoic) acid. These last permeabilities provide a useful tool for the investigation of pH gradients across membranes (Section 3.5).

2.3.2 Ionophore-induced permeability properties of bilayer regions

The high activation energy required to insert an ion into a hydrophobic region is the reason for the extremely low ion permeability of bilayer regions. It follows that if the charge can be delocalized and shielded from the bilayer, the ion permeability might be expected to increase. This is accomplished by a variety of antibiotics synthesized by some micro-organisms, as well as by some synthetic compounds. These are known collectively as ionophores. Ionophores are typically compounds with a molecular weight of 500–2000 and possess a hydrophobic exterior, making them lipid soluble, together with a hydrophilic interior to bind the ion. Ionophores are not natural constituents of energy-conserving membranes, but as tools for investigation they are invaluable.

Ionophores can function as mobile carriers or as channel formers (Fig. 2.2). Mobile carriers diffuse within the membrane, and can typically catalyse the transport of about 1000 ions sec^{-1} across the membrane. They can show an extremely high discrimination between different ions, can work across thick synthetic membranes, and are affected by the fluidity of the membrane. In contrast, channel-forming ionophores discriminate poorly between ions but can be very active, transporting up to 10^7 ions per channel per second.

Ionophores can also be categorized according to the ion transport which they catalyse.

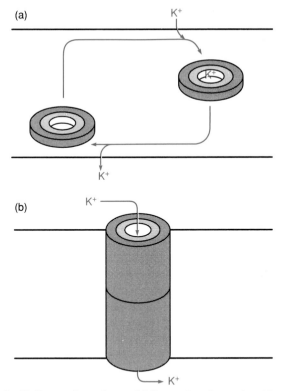

Figure 2.2   Valinomycin and gramicidin catalyse ion uniport by different mechanisms.
Schematic function of (a) valinomycin, a mobile carrier ionophore and (b) gramicidin, a channel-forming ionophore. Note that the ion's hydration sphere is lost in (a) but retained in (b).

2.3.3 Carriers of charge but not protons

(a) *Valinomycin*

Valinomycin (Fig. 2.2) is a mobile carrier ionophore which catalyses the electrical uniport of Cs^+, Rb^+, K^+ or NH_4^+. The ability to transport Na^+ is at least 10^4 less than for K^+. Valinomycin is a natural antibiotic from *Streptomyces* and is a depsipeptide, i.e. it consists of alternating hydroxy and amino acids. The ions lose their water of hydration when they bind to the ionophore. Na^+ cannot be transported because the unhydrated Na^+ ion is too small to interact efficiently with the inward-facing carbonyls of valinomycin, with the result that the complexation energy does not balance that required for the loss of the water of hydration. Because valinomycin is uncharged and contains no ionizable groups, it acquires the charge of the complexed ion. Both the uncomplexed and complexed forms of valinomycin are able to diffuse across the membrane. Therefore, a catalytic amount of ionophore can induce the bulk transport of cations. It is effective in concentrations as low as 10^{-9} M in mitochondria, chloroplasts and synthetic bilayers, and to a more limited extent in bacteria (the outer membrane can exclude it from Gram-negative organisms).

Energy-conserving membranes generally lack a native electrical K^+ permeability and valinomycin can be exploited to induce such a permeability, in order to estimate (Section 4.2) or clamp (Section 4.4) membrane potentials, or to investigate anion transport (Section 2.5).

Other ionophores catalysing K^+ uniport include the enniatins and the nactins (nonactin, monactin, dinactin, etc., so called from the number of ethyl groups in the structure). However, these ionophores do not have such a spectacular selectivity for K^+ over Na^+ as valinomycin.

(b) Gramicidin

Gramicidin is an ionophore which forms transient conducting dimers in the bilayer (Fig. 2.2). Its properties are typical of channel-forming ionophores, with a poor selectivity between protons, monovalent cations and NH_4^+, the ions permeating in their hydrated forms. The capacity to conduct ions is limited only by diffusion, with the result that one channel can conduct up to 10^7 ions s^{-1}.

2.3.4 Carriers of protons but not charge

(a) Nigericin

Nigericin is a linear molecule with heterocyclic oxygen-containing rings together with a hydroxyl group. In the membrane the molecule cyclizes to form a structure similar to that of valinomycin, with the oxygen atoms forming a hydrophobic interior. Unlike valinomycin, nigericin loses a proton when it binds a cation, forming a neutral complex which can then diffuse across the membrane as a mobile carrier. Nigericin is also mobile in its protonated non-complexed form, with the result that the ionophore can catalyse the overall electroneutral exchange of K^+ for H^+ (Fig. 2.3). Other ionophores which catalyse a similar electroneutral exchange include X-537A, monensin, and dianemycin. The latter two show a slight preference for Na^+ over K^+, while X-537A will complex virtually every cation, including organic amines.

Nigericin has been employed to study anion transport (Section 2.5) and to modify the pH gradient across energy-conserving membranes. It is often stated that nigericin abolishes ΔpH across a membrane; in fact the ionophore equalizes the K^+ and H^+ concentration gradients, the final ion gradients depending on the experimental conditions.

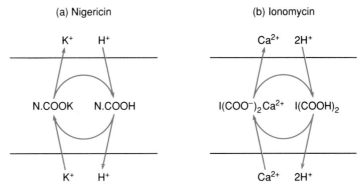

Figure 2.3   Nigericin and ionomycin catalyse electroneutral antiport.
Both (a) nigericin (N) and (b) ionomycin (I) are hydrophobic weak
carboxylic acids permeable across lipid bilayer regions as either the
protonated acid or the neutral salt. Nigericin has a selectivity
$K^+ > Rb^+ > Na^+$ and ionomycin a selectivity $Ca^{2+} > Mg^{2+} \gg Na^+$.

(c) A23187 and ionomycin

A23187 and ionomycin are carboxylic ionophores with a high specificity
for divalent rather than monovalent cations (Fig. 2.3). A23187 catalyses the
electroneutral exchange of Ca^{2+} or Mg^{2+} for $2H^+$ without disturbing
monovalent ion gradients. Ionomycin has a higher selectivity for Ca^{2+} and
has the additional advantage that it is non-fluorescent, allowing its use
in experiments using fluorescent indicators.

2.3.5 Carriers of protons and charge: proton translocators ('uncouplers')

Protonophores, also known as proton translocators or uncouplers, have
dissociable protons and permeate bilayers either as protonated acids or as
the conjugate base, e.g. FCCP (Fig. 2.4). This is possible because these ionophores possess extensive π-orbital systems which so delocalize the charge of
the anionic form that lipid solubility is retained. By shuttling across the membrane they can catalyse the net electrical uniport of protons and increase the
proton conductance of the membrane. In so doing the proton circuit is short-
circuited, allowing the process of Δp generation to be uncoupled from ATP
synthesis. Uncouplers were described long before the formation of the chemiosmotic theory; the demonstration that the majority of these compounds act
by increasing the proton conductance of synthetic bilayers was an important
piece of evidence in favour of the theory and against chemical intermediate
theories which had assigned a specific role to uncouplers in hydrolysing
hypothetical high-energy intermediates.

An indirect proton translocation can be induced in membranes by the combination of a uniport for an ion together with an electroneutral antiport of the

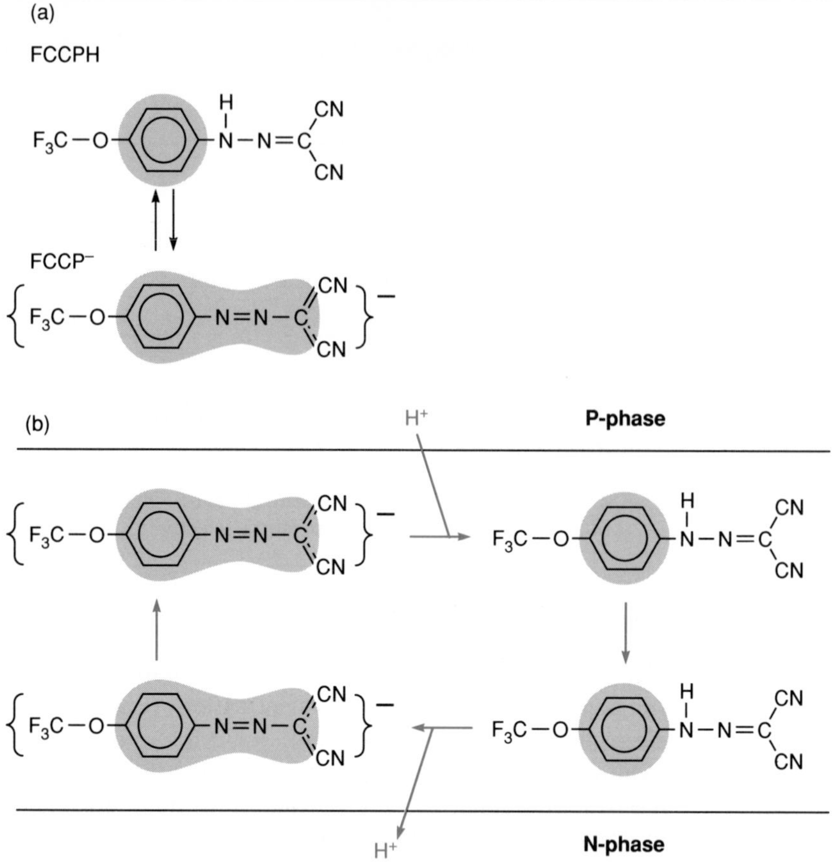

(a)

FCCPH

(b)

Figure 2.4 Protonophores catalyse proton uniport.
Protonophores are lipophilic weak acids permeable across lipid bilayers in
either the protonated or deprotonated forms. (a) FCCP is the most commonly employed example of a protonophore although many such compounds exist. The shaded area represents the extent of the π-orbital
systems. (b) If a Δp exists across the membrane, the protonophore will
cycle catalytically in an attempt to collapse the potential (b). FCCP$^-$ will
be driven to the P-face of the membrane by the membrane potential, while
FCCPH will be driven towards the alkaline or N-phase due to ΔpH. When
sufficient FCCP is present (for most membranes 10^{-9} to 10^{-5} M) the
cycling reduces both $\Delta\psi$ and ΔpH to near zero.

same ion in exchange for a proton. For example, the combination of
valinomycin and nigericin induces a net uniport for H^+, while K^+ cycles
around the membrane. The $Ca^{2+}/2H^+$ ionophores discussed above can also
uncouple mitochondria in the presence of Ca^{2+} since a dissipative cycling is
set up between the native Ca^{2+} uniport (see Section 8.3) and the ionophore.

2.3.6 Lipophilic cations and anions

The ability of π-orbital systems to delocalize charge and enhance lipid solubility has been exploited in the synthesis of a number of cations and anions

which are capable of being transported across bilayer membranes even though they carry charge. Examples include the tetraphenyl phosphonium cation and the tetraphenylborate anion (see Fig. 4.4). These ions are not strictly ionophores, since they do not act catalytically, but are instead accumulated in response to $\Delta\psi$ (Section 3.7.1). Lipophilic cations and anions were of value historically, in demonstrations of their energy-dependent accumulation in mitochondria and inverted sub-mitochondrial particles (SMPs) respectively. These experiments eliminated the possibility of specific cation pumps driven by 'squiggle' (see Fig. 1.9). Subsequently the cations have been employed for the estimation of $\Delta\psi$ (Section 4.2.1).

2.4 PROTEIN-CATALYSED TRANSPORT

The characteristics of protein-catalysed transport across energy-conserving membranes are usually sufficiently distinct from those of bilayer-dependent transport, whether in the absence or presence of ionophores, to make the correct assignment straightforward. The transport proteins of the mitochondrial inner membrane and bacterial cytoplasmic membrane will be discussed in detail in Chapter 8; here we shall merely summarize the distinctions between protein-catalysed and bilayer-mediated transport.

Transport proteins share the features of other enzymes; they can display stereospecificity, can frequently be inhibited specifically, and are genetically determined. This last feature means that it is not possible to make the same kinds of generalizations as for bilayer transport. For example, if FCCP induces proton permeability in mitochondria, it can generally be assumed that the effect will be the same on chloroplasts, bacteria and synthetic bilayers. In contrast, a transport protein may not only be specific to a given organelle but may be restricted to the organelle from one tissue. Thus the citrate carrier is present in liver mitochondria (Section 8.4.7), where it is involved in the export of intermediates for fatty acid synthesis, but is absent from heart mitochondria.

It is sometimes stated that saturation kinetics are characteristic of protein-mediated transport. Although this may be true on occasions, the kinetics of any transport process are so complex, particularly if a membrane potential is applied, that they must be interpreted with care.

The strongest evidence for the involvement of a protein in a transport process is often the existence of specific inhibitors. For example, whereas pyruvate was for many years considered to permeate into mitochondria through bilayers, which is feasible as it is a monocarboxylic weak acid, it was later found that cyanohydroxycinnamate (Section 8.4.9) was a specific transport inhibitor. This provided the first firm evidence for a transport protein for this substrate.

Transport proteins have been studied by many approaches, and this has led to a plethora of names, including carriers, permeases, transporters and translocases, all of which are synonyms for transport protein. The term 'carrier' is particularly inappropriate because there is no evidence that any protein functions by the carrier type of mechanism exemplified by valinomycin.

2.5 THE CO-ORDINATE MOVEMENT OF IONS ACROSS MEMBRANES

The driving forces for the movement of ions across membranes will be derived quantitatively in the next chapter. Here we shall discuss qualitatively how the movements of ions on different carriers within the same membrane may be coupled to each other.

The overriding principle of bulk ion movement across a closed membrane is that charges *must* balance. We have seen that the electrical capacity of, for example, a mitochondrion or bacterium is tiny, and thus that the uncompensated movement of less than 1 nmol of a charged ion per mg protein is sufficient to build up a $\Delta\psi$ of >200 mV. Thus during the operation of a proton circuit the charge imbalance would never exceed this amount, even though the proton current might exceed 1000 nmol H^+ min^{-1} (mg protein)$^{-1}$.

In order to illustrate the coordinate movement of ions we shall first discuss a very simple technique which was much used to establish the pathways and mechanisms of ion transport across the mitochondrial inner membrane: osmotic swelling.

Mitochondria will swell and ultimately burst unless they are suspended in a medium that is isotonic with the matrix. This is why these organelles are frequently handled in buffered 0.25 M sucrose rather than buffer alone. However, mitochondria will also swell if they are placed in a solution in which the principal solute is, unlike sucrose, permeable to the inner membrane. Under such conditions the osmolarity due to the solute becomes equal inside and outside of the mitochondria. Consequently, the osmotic pressure exerted by the contents of the matrix will favour the influx of water and thus mitochondrial swelling will occur. Suspensions of mitochondria are turbid and scatter light. The light scattered is largely a function of the difference in refractive index between the matrix contents and the medium, and any process which decreases this difference will decrease the scattered light. Paradoxically, an increase in the matrix volume due to the influx of a permeable solute results in a decrease in the light scattered as the matrix refractive index approaches that of the medium. This provides a very simple qualitative method for the study of solute fluxes across the mitochondrial inner membrane. Mitochondria are suited for this technique since their matrices can undergo a large increase in volume without bursting, the inner membrane simply unfolding its cristae. 'Swelling' can proceed sufficiently to rupture the outer membrane and release adenylate kinase, which is located in the intermembrane space. Light scattering can be followed either from the decrease in transmitted light in a normal spectrophotometer, or more sensitively by using the $90°$ geometry of a fluorimeter to measure the scattered light directly.

To observe osmotic swelling of mitochondria in ionic media, two criteria must be satisfied:

1. Both the cation and anion of the major osmotic component of the medium must be permeable.
2. The requirement for overall charge balance across the membrane must be respected.

The simplest case to consider is that of mitochondria where both the respiratory chain and the ATP synthase are inhibited. In the example shown in Fig. 2.5 the permeability of rat liver mitochondria for Cl^- and CNS^- is investigated. Mitochondria suspended in either 120 mM KCl or 120 mM KCNS undergo little swelling. However, this does not necessarily demonstrate that the inner membrane is impermeable to these anions, as the K^+ ion is poorly

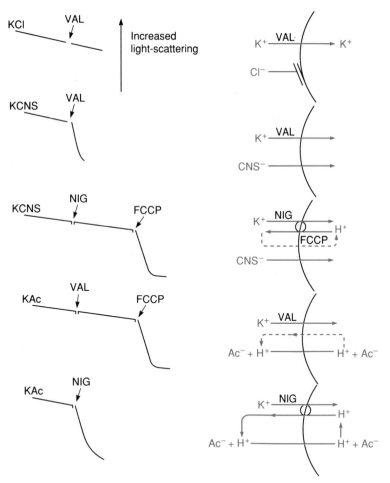

Figure 2.5   Conditions for swelling of non-respiring mitochondria.
For details see text. VAL, valinomycin; NIG, nigericin; Ac^-, acetate (ethanoate) anion.

permeable. To overcome this, an electrical uniport for K^+ can be induced by the addition of valinomycin (Section 2.3.3a) which, being an ionophore, can be confidently predicted to induce the same permeability in the lipid bilayer region of this membrane as it does in synthetic membranes. Rapid swelling is now observed in KCNS, but slow swelling in KCl. It can therefore be concluded that the inner membrane is permeable to CNS^-. The failure of KCl and valinomycin to support swelling does *not* prove that Cl^- is impermeable, for reasons that will now be discussed.

To illustrate a different mode of net solute flux across the membrane, swelling in potassium acetate (potassium ethanoate) is also considered (Fig. 2.5). In order to ensure that there is no rate limitation due to K^+ impermeability, ionophores (valinomycin or nigericin) are present. However, it is clear that the rate of swelling depends on the nature of the permeability induced by the ionophore: nigericin is effective with potassium acetate, while valinomycin is ineffective whilst allowing rapid swelling to occur in KCNS. The reason for the difference is the need for charge balance. K^+ entry catalysed by valinomycin is electrical, while acetate permeates the bilayer as the neutral protonated acid. Therefore, in the presence of valinomycin and potassium acetate a large $\Delta\psi$ (positive inside) rapidly builds up, preventing further K^+ entry. The permeation of acetic acid also ceases, as dissociation of the co-transported proton within the matrix builds up a pH gradient (acid in the matrix) which opposes further acetic acid entry.

These problems are not encountered with potassium acetate in the presence of nigericin. Cation and anion entry are *both* now electroneutral. Also, the proton entering with acetic acid is re-exported by the ionophore in exchange for K^+. It should now be clear that mitochondria swell in KCNS plus valinomycin, but not in KCNS plus nigericin, because permeation of the protonated species HCNS is negligible. It is therefore possible to use swelling not only to determine if a species is permeable but also to determine the mode of entry. Returning to the case of KCl discussed above, swelling does not occur in the presence of either nigericin or valinomycin, thus establishing that Cl^- cannot cross the membrane either by uniport or in symport with H^+.

The requirement, illustrated above and in Fig. 2.5, for swelling that both ions enter by the same mode, be that electrical or electroneutral, can be overcome if an electrical proton uniport is induced by the addition of a proton translocator such as FCCP (Fig. 2.5). Thus with FCCP present swelling occurs in the presence of KCNS plus nigericin or potassium acetate plus valinomycin. The ion fluxes that lead to accumulation of either KCNS or potassium acetate in the matrix are shown in Fig. 2.5.

Matrix volume changes occurring in respiring mitochondria have to take account of the contribution of the protons pumped across the membrane by the respiratory chain. Respiration-dependent swelling occurs in the presence of an electrically permeant cation (which is accumulated due to the membrane potential) and an electroneutrally permeant weak acid (accumulated due to ΔpH; Section 3.5), as shown in Fig. 2.6. Conversely, a rapid contraction of pre-swollen mitochondria occurs on initiation of respiration when the matrix

(a) Optimal swelling

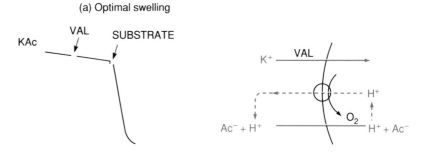

(b) Optimal contraction of pre-swollen mitochondria

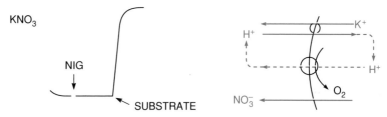

Figure 2.6   Examples of conditions for optimal swelling and contraction of respiring mitochondria.
(a) Mitochondria suspended in potassium acetate. (b) Mitochondria pre-swollen in KNO_3. For details see text.

contains an electroneutrally permeant cation (expelled by ΔpH) and an electrically permeant anion (expelled by $\Delta\psi$).

Consideration of a suspension of phospholipid vesicles for which initially the interior and exterior pH is acidic provides a second example of the requirement for charge balance. If the external pH is subsequently raised into the alkaline range an indicator trapped inside will continue to indicate an acid pH even if the protonophore FCCP is added. Significant proton efflux is not possible because efflux of a tiny quantity of protons generates a membrane potential, positive outside. However, if external K^+ is available and valinomycin is added, protons can efflux via FCCP because the requirement for charge balance is satisfied by the influx of K^+. The internal indicator then signals an alkaline pH.

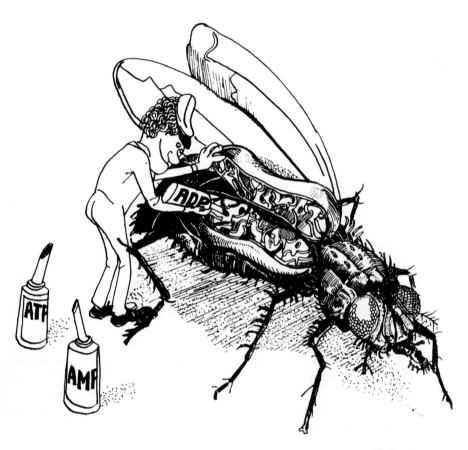

AB TULP

3 QUANTITATIVE BIOENERGETICS: THE MEASUREMENT OF DRIVING FORCES

3.1 INTRODUCTION

Thermodynamics provides the core of bioenergetics, and this chapter is intended to provide an introduction to that part of thermodynamics of specific bioenergetic relevance. We have attempted to de-mythologize some of the more important relationships by deriving them from what we hope are commonsense origins. The reader is strongly advised to follow through the derivations, if only to exorcise the idea, which amazingly still exists in many general biochemistry textbooks, that ATP is a 'high-energy' compound. However, we have also indicated the most essential sections for rapid reference by enclosing them in boxes.

3.1.1 Systems

In thermodynamics three types of system are studied. *Isolated* (or adiabatic) systems are completely autonomous, exchanging neither material nor energy with their surroundings (e.g. a closed insulated vessel). *Closed* systems are materially self-contained, but are capable of exchanging energy across their boundaries (e.g. a hot water bottle). *Open* systems exchange both energy and material with their environment (e.g. all living organisms). The complexity of the thermodynamic treatment of these systems increases as their isolation decreases. Classical equilibrium thermodynamics cannot be applied precisely to open systems because the flow of matter across their boundaries precludes the establishment of a true equilibrium.

All biological systems are open, continually exchanging substrates and end-products with their environment. The most significant contribution of equilibrium thermodynamics to bioenergetics comes from considering individual reactions or groups of reactions as closed systems and asking questions about the nature of the equilibrium state for that reaction. The role of equilibrium thermodynamics is therefore restricted to the following fundamental applications:

(a) Calculating the conditions required for equilibrium in an energy transduction, such as the utilization of the protonmotive force to produce ATP, and by extension determining how far such a reaction is displaced from equilibrium under the actual experimental conditions. It is this displacement from equilibrium which defines the capacity of the reaction to perform useful work.

(b) Eliminating thermodynamically 'impossible' reactions. While no thermodynamic treatment can prove the existence of a given mechanism, equilibrium thermodynamics can readily disprove any proposed mechanism which disobeys its laws. For example, if reliable values are available for the protonmotive force and the free energy for ATP synthesis (concepts which will be developed quantitatively below) it is possible to state unambiguously the lowest value of the H^+/ATP stoichiometry (the number of protons which must enter through the ATP synthase to make an ATP) which would allow ATP synthesis to occur.

(c) Finally, under near-equilibrium conditions, equilibrium parameters can be inserted into the equations of irreversible thermodynamics (which deals with open systems) to give information on the rate of energy flow. This will not be further considered in this book; readers are referred to specialized texts (e.g. Westerhoff and Van Dam, 1987).

3.1.2 Entropy and Gibbs energy change

The universe is by definition an isolated system, and in an isolated system the driving force for a reaction is an increase in entropy, which may be broadly equated to the degree of disorder of the system.

In a closed system a process will occur spontaneously if the entropy of the system *plus its surroundings* increases. Although it is not possible to measure directly the entropy changes in the rest of the universe caused by the energy flow across the boundary of the system, under conditions of constant temperature and pressure this parameter can be calculated from the flow of enthalpy (heat) across the boundaries of the system (Fig. 3.1). The thermodynamic function which takes account of this enthalpy flow is the *Gibbs energy change*, ΔG. ΔG is the quantitative measure of the net driving force (at constant temperature and pressure), and a process which results in a

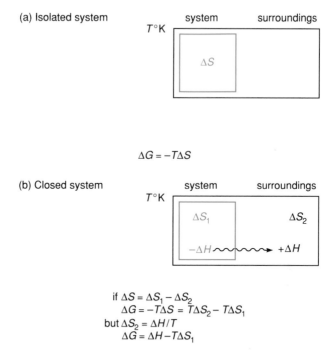

Figure 3.1   Gibbs energy and entropy changes in isolated and closed systems.

ΔG, the Gibbs energy change in a closed system operating at constant temperature and pressure, can be determined using only those parameters which refer to the system itself (the entropy change of the system, ΔS_1, and the enthalpy change, ΔH), while entropy changes in the surroundings need not be determined. A decrease in ΔG in the system implies an increase in ΔS in the universe and so can occur spontaneously (if a pathway exists).

decrease in Gibbs energy ($\Delta G < 0$) is one which causes a net increase in the entropy of the system plus surroundings and is therefore able to occur spontaneously *if* a mechanism is available.

The Gibbs energy change (also termed the free energy change) occurs in bioenergetics in four different guises; indeed the subject might well be defined as the study of the mechanisms by which the different manifestations of Gibbs energy are interconverted.

(1) Gibbs energy changes themselves are used in the description of substrate reactions feeding into the respiratory chain and of the ATP which is ultimately synthesized.
(2) The oxido-reduction reactions occurring in the electron-transfer pathways in respiration and photosynthesis are usually quantified not in terms of Gibbs energy changes but in terms of closely derived redox potential changes.
(3) The available energy in a gradient of ions is quantified by a further variant of the Gibbs energy change, namely the ion electrochemical gradient.

(4) In photosynthetic systems the Gibbs energy available from the absorption of quanta of light can be compared directly with the other Gibbs energy functions.

It should be emphasized that these different conventions merely reflect the diverse historical background of the elements which are brought together in chemiosmotic energy transduction.

3.2 GIBBS ENERGY AND DISPLACEMENT FROM EQUILIBRIUM

Consider a simple reaction A → B, occurring in a closed system. By observing the concentration of reactant A ([A]$_{obs}$) and product B ([B]$_{obs}$) we can calculate the *observed mass action ratio* Γ (capital gamma), equal to [B]$_{obs}$/[A]$_{obs}$. If the mixture of reactant and product happens to be at equilibrium, the mass-action ratio of these equilibrium concentrations [B]$_{equil}$/[A]$_{equil}$ is termed the *equilibrium constant K*. The absolute value of the Gibbs energy (G) increases the further Γ is displaced from K, i.e. the further the reaction is from equilibrium. When Γ/K is plotted logarithmically (Fig. 3.2) a parabola is obtained. The curve shows the following features:

(a) The Gibbs energy content (G) is at a minimum when the reaction is at equilibrium. Thus any change in Γ away from the equilibrium ratio requires an increase in the Gibbs energy content of the system and so cannot occur spontaneously.

(b) The slope of the curve is zero at equilibrium. This means that a conversion of A to B which occurs at equilibrium without changing the mass-action ratio Γ (e.g. by supplying more A and removing B) would cause no change in the Gibbs energy content. Another way of saying this is that the slope ΔG (in units of kJ mol^{-1}) is zero at equilibrium.

(c) When the reaction A → B has not yet proceeded as far as equilibrium, a conversion of A to B without changing the mass action ratio Γ results in a decrease in G, i.e. the slope ΔG is negative. This implies that such an interconversion can occur spontaneously, *provided that a mechanism exists*.

(d) The slope of the curve decreases as equilibrium is approached. Thus ΔG decreases the closer the reaction is to equilibrium. Note that ΔG does *not* equal the Gibbs energy which would be available if the reaction were allowed to run down to equilibrium, but rather gives the Gibbs energy which would be liberated per mole if the reaction proceeded with no change in substrate and product concentrations. This closely reflects the

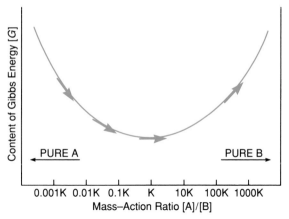

Figure 3.2   Gibbs energy content of a reaction as a function of its displacement from equilibrium.
Consider a closed system containing components A and B at concentrations [A] and [B] which can be interconverted by a reaction [A] $\rightleftharpoons$ [B]. The reaction is at equilibrium when the mass-action ratio [B]/[A] = K, the equilibrium constant. The curve shows qualitatively how the Gibbs energy content (G) of the system varies when the total [A] + [B] is held constant but the mass action ratio is varied away from equilibrium. The tangential arrows represent schematically the Gibbs energy *change*, ΔG, for an interconversion of A to B which occurs at different displacements from equilibrium, *without* changing the mass-action ratio (for example by continuously supplying substrate and removing product).

conditions prevailing *in vivo* where substrates are continuously supplied and products removed.
(e) For the reaction to proceed beyond the equilibrium point would require an input of Gibbs energy, and this therefore cannot occur spontaneously.

The discussion may be generalized and placed on a quantitative footing by considering the reaction where a moles of A and b moles of B react to give c moles of C and d moles of D, i.e.

$$aA + bB \rightleftharpoons cC + dD \qquad [3.1]$$

The equilibrium constant K for the reaction is defined as follows:

$$K = \frac{[C]_{eq}^{c}[D]_{eq}^{d}}{[A]_{eq}^{a}[B]_{eq}^{b}} \text{Molar}^{(c+d-a-b)} \qquad [3.2]$$

where the equilibrium concentration of each component is inserted into the equation to obtain an equilibrium mass-action ratio.
This complicated-looking equation will not be around for long,

however, since we can now define the observed mass-action ratio Γ when the reaction is held away from equilibrium by:

$$\Gamma = \frac{[C]\,^c_{obs}\,[D]\,^d_{obs}}{[A]\,^d_{obs}\,[B]\,^b_{obs}} \quad \text{Molar}^{(c+d-a-b)} \qquad [3.3]$$

Note the symmetry between equations 3.2 and 3.3.

We shall now state, without deriving it, the key equation which relates the Gibbs energy change, ΔG, for the generalized reaction given in equation 3.1 to its equilibrium constant and observed mass-action ratio given in equations 3.2 and 3.3 respectively:

$$\Delta G = -2.3RT \log_{10} K/\Gamma \qquad [3.4]$$

(where the factor 2.3 comes from the conversion from natural logarithms, R is the gas constant, and T is the absolute temperature).

This key equation tells us the following:

- ΔG has a value which is a function of the displacement from equilibrium. The numerical value of the factor $2.3RT$ means that at $25°C$ a reaction which is maintained one order of magnitude away from equilibrium possesses a ΔG of 5.7 kJ mol^{-1}.
- ΔG is negative if $\Gamma < K$ and positive if $\Gamma > K$.

Note that ΔG is a differential, i.e. it measures the change in Gibbs energy which would occur if 1 mol of substrate were converted to product without changing the mass action ratio Γ (e.g. by continuously replenishing substrate and removing product). It does not answer the question 'how much energy is available from running down this reaction to equilibrium'?

3.2.1 ΔG for the ATP hydrolysis reaction

The consequences of equation 3.4 may be illustrated by reference to the hydrolysis of ATP to ADP and P$_i$. At pH 7.0, and in the presence of 10^{-2} M Mg^{2+}, this reaction has an apparent equilibrium constant K' of about 10^5 M. By *apparent* equilibrium constant we mean that obtained by putting the total chemical concentrations of reactants and products into the equation and ignoring the concentration of water, the pH and the effect of ionization state! How such a surprising oversimplification is possible will be explained later. The equation for the equilibrium of the ATP hydrolysis reaction is:

$$K' = \frac{[\Sigma ADP]\,[\Sigma Pi]}{[\Sigma ATP]} = 10^5 \text{ M} \qquad [3.5]$$

where each concentration represents the total sum of the concentrations of the

Table 3.1 The Gibbs energy change for the hydrolysis of ATP to ADP + Pi as a function of the displacement from equilibrium

If $K' = 1 \times 10^5$ M at pH 7 and $[Mg^{2+}] = 10^{-2}$ M and if $[Pi] = 10^{-2}$ M			
Γ' (M)	K'/Γ'	ΔG (kJ mol^{-1})	[ATP/ADP]
10^5	1	0	10^{-7} (equilibrium)
10^3	10^2	-11.4	10^{-5}
10	10^4	-22.8	10^{-3}
1	10^5	-28.5	10^{-2} ('standard conditions')
10^{-1}	10^6	-34.2	10^{-1}
10^{-3}	10^8	-45.6	10
10^{-5}	10^{10}	-57	10^3 (in cytoplasm)

different ionized species of each component, including that complexed to Mg^{2+} (see below).

As equilibrium is attained when Γ is 10^5 M, the equilibrium concentration of ATP in the presence of 10^{-2} M Pi and 10^{-3} M ADP (which are approximate figures for the cytoplasm) would be only 10^{-10} M, or about one part per ten million of the total adenine nucleotide pool!

The variations of ΔG with the displacement of the ATP hydrolysis mass-action ratio from equilibrium are shown in Table 3.1. Mitochondria are able to maintain a mass-action ratio in the incubation medium which is as low as 10^{-5} M, ten orders of magnitude away from equilibrium. Under these conditions the incubation might contain 10^{-2} M Pi, 10^{-2} M ATP and only 10^{-5} M ADP. To synthesize ATP under these conditions requires an input of Gibbs energy of 57 kJ per mole of ATP produced.

Note that ΔG for ATP *synthesis* (often referred to as the 'phosphorylation potential', ΔG_p) is obtained from the corresponding value for ATP hydrolysis by simply changing the sign.

3.2.2 The uses and pitfalls of standard Gibbs energy, ΔG^0

A special case of the general equation for ΔG (equation 3.4) occurs under the totally hypothetical condition when the concentration of all reactants and products are in their 'standard states', i.e. 1 M for solutes, a pure liquid or a pure gas at 1 atmosphere. These conditions define the standard Gibbs energy change ΔG^0.

Considering again our generalized reaction in equation 3.1, under these 'standard' conditions Γ has a value of 1 M$^{(c+d-a-b)}$ and equation 3.4 reduces to:

$$\Delta G^0 = -2.3RT \log_{10} K \qquad [3.6]$$

(note that the term $\log_{10} K$ is dimensionless because the units of K are cancelled by those of Γ).

Equation 3.6 is frequently misunderstood. It is important to appreciate that ΔG^0 is essentially the logarithm of the equilibrium constant and as such gives in itself no information whatsoever concerning the Gibbs energy of the reaction in the cell. It is therefore absolutely incorrect to use ΔG^0 values to predict whether a reaction can occur spontaneously or to estimate the Gibbs energy available from a reaction.

Equation 3.6 can, however, be used to derive a form of the Gibbs energy equation in which the equilibrium constant is substituted by ΔG^0. If we take Eq. 3.4 and divide both K and Γ by the standard state concentrations to make them dimensionless, and then rearrange the equation, we get:

$$\Delta G = -2.3RT \log_{10} K + 2.3RT \log_{10} \Gamma \qquad [3.7]$$

Combining with Eq. 3.6 and eliminating K gives:

$$\Delta G = \Delta G^0 + 2.3RT \log_{10} \Gamma \qquad [3.8]$$

Equation 3.8 is the most common form of the Gibbs energy equation and the one found in most textbooks. Just as Eq. 3.4 has terms for Γ and K, so Eq. 3.8 has terms for Γ and ΔG^0. Equation 3.8 expresses the important fact that the Γ term as well as ΔG^0 determine the Gibbs energy under any particular set of concentrations of reactants and products found in a cell. Note that Eq. 3.8 reverts to Eq. 3.6 at equilibrium when $\Delta G = 0$ and, of course, $\Gamma = K$. Both Eq. 3.4 and Eq. 3.8 can be used correctly to calculate ΔG; however Eq. 3.4 is more intuitive since it emphasizes the fact that ΔG is a function of the extent to which a reaction is removed from equilibrium.

Finally, from Eq. 3.8 it is not immediately evident that ΔG^0 and $2.3RT \log_{10} \Gamma$ are dimensionally homogeneous terms or why apparent equilibrium constants and apparent mass-action ratios (see below) can be used that make simplifying assumptions about the states of ionization of reactants and products, the pH etc.

3.2.3 Absolute and apparent equilibrium constants and mass action ratios

To avoid confusion or ambiguity in the derivation of equilibrium constants, and hence Gibbs energy changes, a number of conventions have been adopted. Those most relevant to bioenergetics are the following:

(a) True thermodynamic equilibrium constants (K) are defined in terms of the chemical activities rather than the concentrations of the reactants and products. Generally in biochemical systems it is not possible to determine the activities of all the components, and so equilibrium constants are calculated from concentrations. This introduces no error as long as the observed mass-action ratio and the equilibrium constants are calculated under comparable conditions (remember ΔG is calculated from the ratio of Γ and K).

(b) When water appears as either a reactant or product in dilute solutions, it is considered to be in its standard state (which is the pure liquid at 1 atmosphere) under both equilibrium and observed conditions. This means that the water term can be omitted from both the equilibrium and observed mass-action ratio equations (see, for example, Eq. 3.5).

(c) If one or more of the reactants or products are ionizable, or can chelate a cation, there is an ambiguity as to whether the equilibrium constant should be calculated from the total sum of the concentrations of the different forms of a compound, or just from the concentration of that form which is believed to participate in the reaction. The hydrolysis of ATP to ADP and Pi is a particularly complicated case: not only are all the reactants and products partially ionized at physiological pH, but also Mg^{2+}, if present, chelates ATP and ADP with different affinities. Thus ATP can exist at pH 7 in the following forms:

$$[\Sigma ATP] = [ATP^{4-}] + [ATP^{3-}] + [Mg.ATP^{2-}] + [Mg.ATP^-] \quad [3.9]$$

If it were known that the true reaction was:

$$Mg.ATP^{2-} + H_2O \rightleftharpoons Mg.ADP^- + HPO_4^{2-} + H^+ \quad [3.10]$$

then the true equilibrium constant would be:

$$K = \frac{[Mg.ADP^-][HPO_4^{2-}][H^+]}{[Mg.ATP^{2-}]} M^2 \quad [3.11]$$

This equilibrium constant would be independent of pH or Mg^{2+}, as these factors are allowed for in the equation. However, the reacting species are *not* known unambiguously, and even if they were, their concentrations would be difficult to assay, as enzymatic or chemical assay determines the total concentration of each compound (e.g. ΣATP).

In practice, therefore, an apparent equilibrium constant, K', is employed, calculated from the total concentrations of each reactant and product, ignoring any effects of ionization or chelation and omitting any protons which are involved (see Eq. 3.5).

The most important limitation of the apparent equilibrium constant is that K' is not a universal constant, but depends on all those factors which are omitted from the equation, such as pH and cation concentration. K' is thus only valid for a given pH and cation concentration, and must be qualified by information about these conditions. As the standard Gibbs energy change is derived directly from the equilibrium constant, this parameter must be similarly qualified. Finally, and most importantly, the apparent mass action ratio, Γ', must be calculated under exactly the same set of assumptions; if this is done, when the ratio K'/Γ' is calculated for Eq. 3.4 all the assumptions cancel out and a true and meaningful ΔG is obtained. In biochemistry the terms $\Delta G^{0'}$ and K' are frequently used to specify that a $[H^+]$ of 10^{-7} M is being considered, but in principle these parameters can be specified for any condition of pH, ionic strength, temperature, $[Mg^{2+}]$ etc. that is convenient − as long as Γ' is *always* calculated under exactly the same set of conditions.

3.2.4 **The myth of the 'high-energy phosphate bond'**

It is frequently, and misleadingly, supposed that the phosphate anhydride bonds of ATP are 'high-energy' bonds which are capable of storing energy and driving reactions in otherwise unfavourable directions. However, it should be clear from Table 3.1 that it is the extent to which the observed mass-action ratio is displaced from equilibrium which defines the capacity of the reactants to do work, rather than any attribute of a single component. A hypothetical cell could utilize any reaction to transduce energy from the mitochondrion. For example, if the glucose-6-phosphatase reaction were maintained ten orders of magnitude away from equilibrium, then glucose 6-phosphate would be just as capable of doing work in the cell as is ATP. Conversely, the Pacific Ocean could be filled with an equilibrium mixture of ATP, ADP and Pi, but the ATP would have no capacity to do work.

3.3 OXIDATION–REDUCTION (REDOX) POTENTIALS

3.3.1 Redox couples

Both the mitochondrial respiratory chain and the photosynthetic electron-transfer chains operate as a sequence of reactions in which electrons are transferred from one component to another. While many of these components simply gain one or more electrons in going from the oxidized to the reduced form, in others the gain of electrons induces an increase in the pK of one or more ionizable groups on the molecule, with the result that reduction is accompanied by the gain of one or more protons.

Cytochrome c undergoes a one-electron reduction:

$$Fe^{3+}.cyt\ c + 1e^- \rightleftharpoons Fe^{2+}.cyt\ c \qquad [3.12]$$

NAD^+ undergoes a two-electron reduction and gains one proton:

$$NAD^+ + 2e^- + H^+ \rightleftharpoons NADH \qquad [3.13]$$

while ubiquinone (UQ) undergoes a two-electron reduction followed by the addition of two protons:

$$UQ + 2e^- + 2H^+ \rightleftharpoons UQH_2 \qquad [3.14]$$

This last is often, but inaccurately, referred to as a 2H-transfer.

Redox reactions are not restricted to the electron-transport chain. For example, lactate dehydrogenase also catalyses a redox reaction:

$$\text{Pyruvate} + NADH + H^+ \rightleftharpoons \text{Lactate} + NAD^+ \qquad [3.15]$$

While all oxidation–reduction reactions can quite properly be described in thermodynamic terms by their Gibbs energy changes, since the reactions involve the transfer of electrons electrochemical parameters can be employed. Although the thermodynamic principles are the same as for the Gibbs energy change, the origins of oxido-reduction potentials in electrochemistry sometimes obscure this relationship.

The additional facility afforded by an electrochemical treatment of a redox reaction is the ability to dissect the overall electron transfer into two half-reactions, involving respectively the donation and acceptance of electrons. Thus Eq. 3.15 can be considered as the sum of two half-reactions:

$$NADH \rightleftharpoons NAD^+ + H^+ + 2e^- \qquad [3.16]$$

and

$$\text{Pyruvate} + 2H^+ + 2e^- \rightleftharpoons \text{Lactate} \qquad [3.17]$$

(note that these add together to give Eq. 3.15). A reduced–oxidized pair such as $NADH/NAD^+$ is termed a redox couple.

3.3.2 Determination of redox potentials

Each of the half-reactions above is reversible, and so can in theory be described by an equilibrium constant. However, it is not immediately apparent how to treat the electrons, which have no independent existence in solution. A similar problem is encountered in electrochemistry when investigating the equilibrium between a metal (i.e. the reduced form) and a solution of its salt (i.e. the oxidized form). In this case the tendency of the couple to donate electrons is quantified by forming an electrical cell from two half-cells, each consisting of a metal electrode in equilibrium with a 1 M solution of its salt. An electrical circuit is completed by a bridge which links the solutions without allowing them to mix. The electrical potential difference between the electrodes may then be determined experimentally.

To facilitate comparison, electrode potentials are expressed in relation to the standard hydrogen electrode:

$$2H^+ + 2e^- \rightleftharpoons H_2 \qquad [3.18]$$

Hydrogen gas at 1 atmosphere is bubbled over the surface of a platinum electrode which has been coated with finely divided platinum to increase the surface area. When this electrode is immersed in 1 M H^+ the absolute potential of the electrode is defined as zero (at 25°C). The standard electrode potential of any metal/salt couple may now be determined by forming a cell comprising the unknown couple together with the standard hydrogen electrode, or more conveniently with secondary standard electrodes whose electrode potentials are known.

A similar approach has been adopted for biochemical redox couples. As with the hydrogen electrode it is not feasible to construct an electrode out of the reduced component of the couple, so a platinum electrode is employed. However, unlike the metal/salt and H_2/H^+ couples, both components can generally exist in aqueous solution, and standard conditions are defined in which both the oxidized and reduced components are present at unit activity, or 1 M in concentration terms, including any protons, i.e. pH = 0. Note the parallel to the conditions for ΔG^0. The experimentally observed potential relative to the hydrogen electrode is termed the standard redox potential, E^0.

In only a few cases of bioenergetic relevance do the oxidized and reduced components of the couple (e.g. the oxidized and reduced states of a cytochrome) equilibrate with the Pt electrode sufficiently rapidly for a stable potential to be registered. In most cases a low concentration of a second redox couple, capable of reacting with both the primary redox couple (e.g. the cytochrome) and the platinum electrode, is added to act as a redox mediator. As will be shown below, the primary redox couple and the redox mediator achieve equilibrium when they exhibit the same redox potential. As long as the concentration relationships of the primary couple are not disturbed, the electrode can therefore register the potential of the primary couple. The use of one or more mediating couples is of particular importance when the redox potentials of membrane-bound components are being investigated (Section 5.4.1).

3.3.3 Redox potential and [oxidized]/[reduced] ratio

Just as the standard Gibbs energy change ΔG^0 does not reflect the actual conditions existing in the cell, the standard redox potential E^0 must be qualified to take account of the relative concentrations of the oxidized and reduced species.

The actual redox potential E at pH = 0 for the redox couple:

$$ox + ne^- \rightleftharpoons red$$

is given by the relationship:

$$E = E^0 + \frac{2.3RT}{nF} \log_{10}\left(\frac{[ox]}{[red]}\right) \qquad [3.19]$$

where R is the gas constant and F the Faraday constant. Note that this equation is closely analogous to the 'conventional' equation involving standard Gibbs energy changes (Eq. 3.8).

3.3.4 Redox potential and pH

In many cases (e.g Eqs 3.13 and 3.14) protons are involved in the redox reaction, in which case the generalized half-reaction becomes:

$$ox + ne^- + mH^+ \rightleftharpoons red$$

The standard redox potential at a pH other than zero becomes more negative than E^0 with a dependency of $2.3RT/F(m/n)$ mV per pH unit. This corresponds to -60 mV/pH when $m = n$ and -30 mV/pH when $m = 1$ and $n = 2$ (Table 3.2). Note that the potentials are still calculated relative to the standard hydrogen electrode at pH $= 0$.

The usual biochemical convention is to define redox potentials for pH 7. The standard redox potential under these conditions is given the symbol $E^{0\prime}$ and is also referred to as the mid-point potential $E_{m,7}$ because inspection of Eq. 3.19 shows that it is also the potential where the concentrations of the oxidized and reduced forms are equal.

Note that although the potential of the standard hydrogen electrode always remains zero, $E_{m,7}$ for the $H^+/\frac{1}{2}H_2$ couple is $7 \times (-60) = -420$ mV.

The actual redox potential at a pH of x ($E_{h,pH = x}$) is related to the mid-point potential at that pH by the relationship:

$$E_{h,pH = x} = E_{m,pH = x} + \frac{2.3RT}{F} \log_{10}\left(\frac{[ox]}{[red]}\right) \qquad [3.20]$$

Table 3.2   Some mid-point potentials

				Change in E_m (mV) when
$ox + ne^- + mH^+ = red$				
	n	m	$E_{m,7}$ (mV)	pH increased by 1 unit
Methyl viologen ox/red	1	0	-450	0
Ferredoxin ox/red	1	0	-430	0
$H^+/\frac{1}{2}H_2$ (H_2 1 atm)	1	1	-420	-60
$NAD^+/NADH$	2	1	-320	-30
$NADP^+/NADPH$	2	1	-320	-30
Menaquionone/menaquinol	2	2	-74	-60
Fumarate/succinate	2	2	$+30$	-60
Ubiquinone/ubiquinol	2	2	$+40$	-60
Ascorbate ox/red	2	1	$+60$	-30
PMS ox/red	2	1	$+80$	-30
Cyt c ox/red	1	0	$+220$	0
DCPIP/DCPIPH$_2$	2	2	$+220$	-60
DAD/DADH$_2$	2	2	$+275$	-60
TMPD ox/red	1	0	$+260$	0
Ferricyanide ox/red	1	0	$+420$	0
O_2(1 atm)/2H_2O (55 M)	4	4	$+820$	-60

DAD is 2,3,5,6-tetramethylphenylene diamine; PMS is phenazine methosulphate; TPMD is N,N,N',N'-tetramethyl-p-phenylene diamine; DCPIP is 2,6-dichlorophenolindophenol. For further details see Prince et al. (1981).

The characteristic variation of E_h with the ratio of oxidized to reduced component is shown in Fig. 3.3.

3.3.5 Redox potential difference and the relation to ΔG

The ΔG which is available from a redox reaction is proportional to the difference in the actual redox potentials ΔE_h between the donor and acceptor redox couples. (Note that the difference in redox potential between two couples (redox potentials $E_{(A)}$ and $E_{(B)}$) is written in most books, as E, but we believe that use of ΔE_h clarifies that a *difference* between two couples, or a redox span in an electron-transport system, is being considered).

In the case of the mitochondrion, $E_{h,7}$ for the NADH/NAD$^+$ couple is about -280 mV and $E_{h,7}$ for the O_2/H_2O couple is $+780$ mV (note these values differ slightly from the mid-point potentials shown in Table 3.2 because of the relative concentrations of the NADH and NAD$^+$ and because oxygen comprises only 20% of air). The redox potential difference $\Delta E_{h,7}$ of 1.16 V is the measure of the thermodynamic disequilibrium between the couples. In general terms for the redox couples A and B:

$$\Delta E_h = E_{h(A)} - E_{h(B)} \qquad [3.21]$$

A final note about nomenclature: a powerful oxidizing agent is an oxidized component of a redox couple with a relatively positive E_m (e.g O_2), whereas a powerful reducing agent is the reduced component of a redox couple with a relatively negative E_m (e.g NADH).

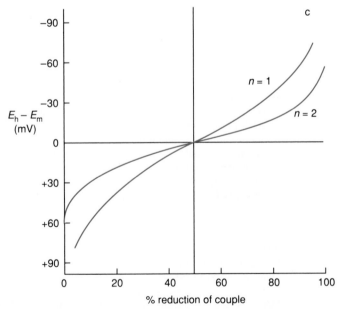

Figure 3.3 The variation of E_h with the extent of reduction of a redox couple.
$n = 1$, $n = 2$ refer to to one- and two-electron oxido-reductions respectively.

3.4 ION ELECTROCHEMICAL POTENTIAL DIFFERENCES

We have tried to emphasize in this chapter that the Gibbs energy change in a process is a function of its displacement from equilibrium. The disequilibrium of an ion or metabolite across a membrane can be subjected to the same quantitative treatment. As before, the derivation is not only valid for energy-transducing membranes, but has equal applicability to all membrane transport processes.

There are two forces acting on an ion gradient across a membrane, one due to the concentration gradient of the ion and one due to the electrical potential difference between the aqueous phases separated by the membrane (the 'membrane potential' $\Delta\psi$). These can initially be considered separately.

Consider the Gibbs energy change for the transfer of 1 mol of solute across a membrane from a concentration $[X]_A$ to a concentration $[X]_B$ where the volumes of the two compartments are sufficiently large that the concentrations do not change significantly.

In the absence of a membrane potential ΔG is given by:

$$\Delta G(\text{kJ mol}^{-1}) = 2.3RT \, \log_{10}\left(\frac{[X]_B}{[X]_A}\right) \tag{3.22}$$

Note that this equation is closely analogous to that for scalar reactions (Eq. 3.4). In particular ΔG in both cases is 5.7 kJ mol^{-1} for each tenfold displacement from equilibrium.

The second special case is for the transfer of an ion driven by a membrane potential in the absence of a concentration gradient. In this case ΔG when 1 mol of anion X^{m+} is transported down an electrical potential of $\Delta\psi$ mV is given by:

$$\Delta G \, (\text{kJ mol}^{-1}) = -mF \, \Delta\psi \tag{3.23}$$

where F is the Faraday constant. In the general case, the ion will be affected by *both* concentration *and* electrical gradients, and the net ΔG when 1 mol of X^{m+} is transported down an electrical potential of $\Delta\psi$ mV from a concentration of $[X^{m+}]_A$ to $[X^{m+}]_B$ is given by the general electrochemical equation:

$$\Delta G \, (\text{kJ mol}^{-1}) = -mF \, \Delta\psi + 2.3RT \, \log_{10}\left(\frac{[X^{m+}]_B}{[X^{m+}]_A}\right) \tag{3.24}$$

ΔG in this equation is often expressed as the ion electrochemical gradient $\Delta\bar{\mu}_{X^{m+}}$ (kJ mol^{-1}).

In the specific case of the proton electrochemical gradient, $\Delta\bar{\mu}_{H^+}$, Eq. 3.24 can be considerably simplified since pH is a logarithmic function of $[H^+]$:

$$\Delta\bar{\mu}_{H^+} = -F \, \Delta\psi + 2.3RT \, \Delta\text{pH} \tag{3.25}$$

where pH is defined as (pH in the P-phase) − (pH in the N-phase).

Mitchell defined the term *protonmotive force* (Δp), where:

$$\Delta p(\text{mV}) = -(\Delta \bar{\mu}_{H^+})/F \qquad [3.26]$$

This facilitates a thermodynamic comparison with the redox potential differences in the electron-transfer chain complexes which generate the proton gradient, as well as emphasizing that we are dealing with a *potential* driving a proton circuit. A $\Delta \bar{\mu}_{H^+}$ of 1 kJ mol^{-1} corresponds to a Δp of 10.4 mV. Using Δp and substituting values for R and T at 25°C the final equation is:

$$\Delta p(\text{mV}) = \Delta \psi - 59\Delta \text{pH} \qquad [3.27]$$

$\Delta \psi$ is defined as P-phase − N-phase.

3.5 PHOTONS

In photosynthetic systems, the primary source of Gibbs energy is the quantum of electromagnetic energy, or photon, which is absorbed by the photosynthetic pigments. The energy in a single photon is given by $h\nu$, where h is Planck's constant (6.62×10^{-34} J s) and ν is the frequency of the radiation (s^{-1}). One photon interacts with one molecule, and therefore N photons, where N is Avogadro's constant, will interact with 1 mol.

The energy in 1 mol (or einstein) of photons is therefore:

$$\Delta G = Nh\nu = Nhc/\lambda = 120\,000/\lambda \text{ kJ einstein}^{-1} \qquad [3.28]$$

where c is the velocity of light, and λ is the wavelength in nanometres. Even the absorption of an einstein of red light (600 nm) makes available 200 kJ mol^{-1} which compares favourably with the Gibbs energy changes encountered in bioenergetics.

3.6 BIOENERGETIC INTERCONVERSIONS AND THEIR STOICHIOMETRIES

The critical stages of chemiosmotic energy transduction involve the interconversions of ΔG between the different forms discussed in the previous sections. In the case of the mitochondrion these are: redox potential difference (ΔE_h) → protonmotive force, (Δp) → ΔG for ATP synthesis (the 'phosphorylation potential' ΔG_p).

While bioenergetic systems are open *in vivo*, i.e. they operate under non-equilibrium conditions, with isolated organelles it is frequently possible to allow a given interconversion to achieve a true equilibrium by the simple expedient of inhibiting subsequent steps. For example, isolated mitochondria can achieve an equilibrium between Δp and ΔG_p if reactions which hydrolyse ATP are absent.

In some cases a true equilibrium is not possible in practice. For example, because of the inherent proton permeability of the mitochondrial membrane (Section 4.6) it is not possible to achieve a true equilibrium between the respiratory chain ΔE_h and Δp, since there is always some net leakage of protons across the membrane which results in the steady-state value of Δp lying below its equilibrium value with ΔE_h. Consequently there is always some flux of electrons from NADH to oxygen. Under these conditions it is valid to obtain a number of values for different flux rates and to extrapolate back to the static head condition of zero flux.

A test of whether an interconversion is at equilibrium is to establish whether a slight displacement in conditions will cause the reaction to run in reverse. In respiring mitochondria this test can be fulfilled by the ATP synthase and by two of the three respiratory chain proton pumps (complexes I and III, see Chapter 5). A process is, of course, at equilibrium when the overall ΔG is zero.

3.6.1 The relation between ΔG and E_h

There is a simple and direct relationship between the redox potential difference of two couples, ΔE_h, and the Gibbs energy change ΔG accompanying the transfer of electrons between the couples:

$$\Delta G = -nF \, \Delta E_h \qquad [3.29]$$

where n is the number of electrons transferred, and F is the Faraday constant. From this it is apparent that an oxido-reduction reaction is at equilibrium when $\Delta E_h = 0$. Table 3.3 relates redox potential differences and Gibbs energy changes. Note that every mole of electrons transferred down the respiratory chain from NADH to O_2 can yield over 100 kJ of Gibbs energy.

One complication arises where the donor and acceptor redox couples are on the opposite sides of a membrane across which an electrical potential exists (Fig. 3.4). Examples of this occur in the mitochondrial respiratory chain

Table 3.3   Interconversion between redox potential difference and Gibbs energy change for one-electron and two-electron transfers

	ΔG (kJ mol^{-1})	
ΔE_h (mV)	$n = 1$	$n = 2$
0	0	0
+ 100	− 9.6	− 19.3
+ 200	− 19.3	− 38.6
+ 500	− 48.2	− 96.5
+ 1000	− 96.5	− 193
+ 1200	− 116	− 231

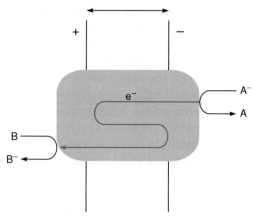

Figure 3.4 **ΔE_h and ΔG for an electron transfer between redox couples located on opposite sides of a membrane sustaining a membrane potential.**

$$\Delta E_h = (E_A - E_B)$$
$$\Delta G = -nF(\Delta E_h + \Delta \psi)$$

(Chapter 5). If the electron enters from the negative compartment and is transferred to a couple in the positive compartment, then it is intuitive that the process will yield more energy than if there was no $\Delta \psi$ to provide an extra 'push' for the electron. The value of the membrane potential must be added to the redox potential difference to calculate the effective Gibbs energy change:

$$\Delta G = -nF(\Delta E_h + \Delta \psi) \qquad [3.30]$$

3.6.2 Proton pumping by electron-transfer chains: ΔE_h and Δp

If two electrons passing through a redox span ΔE_h within the respiratory chain pump n protons across the membrane against a protonmotive force of Δp then equilibrium would be attained when:

$$n \, \Delta p = 2\Delta E_h \qquad [3.31]$$

Thus the higher the H^+/O stoichiometry (n) of a respiratory chain complex with a particular value of ΔE_h, the *lower* the equilibrium Δp which can be attained, just as a bicycle has less ability to climb a hill in high rather than low gear.

Note that Eq. 3.31 only holds if the two electrons enter and leave the respiratory span on the same side of the membrane. If, as in the case of electron transfer from succinate dehydrogenase (on the matrix face) to cytochrome c (on the cytoplasmic face) the electrons effectively cross the membrane they will be aided by the membrane potential (Fig. 3.4) and the relationship becomes:

$$n \, \Delta p = 2(\Delta E_h + \Delta \psi) \qquad [3.32]$$

3.6.3 Equilibrium at the ATP synthase: Δp and ΔG_p

The equilibrium relationship between Δp and the phosphorylation potential in the N-phase, e.g. the mitochondrial matrix, $\Delta G_{p,matrix}$, is given by:

$$\Delta G_{p,matrix} = n'F\Delta p \qquad [3.33]$$

where n' is the H^+/ATP stoichiometry. Note that the higher the H^+/ATP stoichiometry (n') the *higher* the ΔG_p which can be attained at equilibrium at any particular value of Δp.

Since one additional proton is expended in the overall transport of Pi and ADP into, and of ATP out of, the mitochondrial matrix (Section 8.4.3) the relationship for the eukaryotic cytoplasmic phosphorylation potential $\Delta G_{p,cyto}$ becomes:

$$\Delta G_{p,cyto} = (n' + 1) \, \Delta p \qquad [3.34]$$

As n' is either 2 or 3 (Section 4.5) this means that a substantial proportion of the Gibbs energy for the cytoplasmic ATP/ADP + Pi pool comes from the transport step rather than the ATP synthase itself. Naturally this occurs at a cost: the overall H^+/ATP stoichiometry is increased and the $ATP/2e^-$ (Sections 4.5 and 4.10.2) is decreased.

3.6.4 Thermodynamic stoichiometries

Since Eqs 3.31, 3.33 and 3.34 each contain a term for the stoichiometry, it is possible to determine the thermodynamic parameters at equilibrium, substitute these values into the equations and hence calculate the stoichiometry term, without actually measuring the movement of protons across the membrane. This is known as the thermodynamic stoichiometry. Naturally such a calculation is only as accurate as the determination of the thermodynamic parameters, but it does offer an alternative approach to the non-steady-state technique which will be discussed in Chapter 4.

While it might be thought that the equations of thermodynamic equilibrium would offer a simple and unambiguous determination of stoichiometry, in fact considerable controversy has arisen over the interpretation of the results. The questions which have been debated include the following:

- Does the measured Δp give an accurate measure of the proton electro-chemical potential driving ATP synthesis, or is there a 'localized' proton circuit involving an intimate coupling between a respiratory chain complex and an ATP synthase complex, which is out of equilibrium with the 'bulk-phase' protons? (This will be discussed in Section 4.9.)
- Is the stoichiometry constant or do the proton-translocating complexes 'change gear' at high Δp (Section 4.9)?

Both these points will be discussed in Chapter 4.

3.6.5 The 'efficiency' of oxidative phosphorylation

A statement of the type 'oxidation of NADH by O_2 has a $\Delta G^{0\prime}$ of -220 kJ mol^{-1} whilst ATP synthesis has a $\Delta G^{0\prime}$ of $+31$ kJ mol^{-1}, and thus if three ATP molecules are synthesized for each NADH oxidized, mitochondrial oxidative phosphorylation traps approximately 93 kJ mol^{-1} of the energy available from NADH oxidation, an efficiency of 42% appears in many textbooks of biochemistry. This analysis has at least two shortcomings. First, it refers to standard conditions which, recall, refer to 1 M solutions and 1 atmos. O_2 that are not found in cells. Second, there is no basis in physical chemistry for dividing an output ΔG (93 kJ mol^{-1} in this case) by the input ΔG (220 kJ mol^{-1}) to calculate an efficiency (Cornish-Bowden, 1983).

Under cellular conditions 2 mol of electrons flowing from the NADH/NAD$^+$ couple to oxygen liberate about 220 kJ (a value that, atypically of reactions in cells, happens to be close to the standard state value) and in the ideal case when there is no proton leak across the membrane this would be conserved in the generation of a Δp of some 200 mV, while about 10 mol H$^+$ are pumped across the membrane. The energy initially conserved in the proton gradient is thus about $10 \times 200 \times F = 200$ kJ. If 3 mol ATP are synthesized per pair of electrons passing down the respiratory chain and the ATP is subsequently exported to the cytoplasm at a ΔG_p of about 60 kJ mol^{-1}, i.e. 180 kJ, then it can be seen that oxidative phosphorylation machinery can closely approach equilibrium and that there are no large energy losses between electron transport and ATP synthesis. In this sense the machine can be regarded as highly efficient or effective. However, it is important to realize that as ATP turnover increases, e.g. in an exercising muscle, the ΔG_p will be significantly lower than 60 kJ mol^{-1}. Under these conditions the overall ΔG is more negative but this is just the inevitable price of running a reaction away from close-to-equilibrium conditions.

Comparison of oxidative phosphorylation in mitochondria with that in *E. coli* (Section 5.13.2) shows that in the latter NADH oxidation is coupled to the synthesis of fewer ATP molecules than in mitochondria. The reason for this appears to be that fewer protons are pumped for each pair of electrons flowing from NADH to oxygen. As all other energetic parameters (ΔG and Δp) are similar it could be said that oxidative phosphorylation is less 'efficient' in the bacterium owing to failure to conserve fully the energy from respiration in the form of the Δp.

In practice all energy-transducing membranes have a significant proton leak and thus the actual output is reduced so that equilibrium between ATP synthesis and respiration is not reached. Irreversible thermodynamics, which is beyond the scope of this book, is able to calculate that the true efficiency (i.e. power output divided by power input) (Stucki, 1983) is optimal when the mitochondria are synthesizing ATP rapidly, since the proton leak is greatly decreased (Section 4.7) and most proton flux is directed through the ATP synthase.

3.7 EQUILIBRIUM DISTRIBUTIONS OF IONS, WEAK ACIDS AND WEAK BASES

3.7.1 Charged species and $\Delta\psi$

As with all Gibbs energy changes, an ion distribution is at equilibrium across a membrane when ΔG is zero, and hence when $\Delta\tilde{\mu}_{X^{m+}} = 0$. Under these conditions the ion electrochemical equation (Eq. 3.24) becomes:

$$0 = -mF\,\Delta\psi + 2.3RT\,\log_{10}\left(\frac{[X^{m+}]_B}{[X^{m+}]_A}\right) \qquad [3.35]$$

This rearranges to give the Nernst equation, relating the equilibrium distribution of an ion to the membrane potential:

$$\Delta\psi = \frac{2.3RT}{mF}\log_{10}\left(\frac{[X^{m+}]_B}{[X^{m+}]_A}\right) \qquad [3.36]$$

An ion can thus come to electrochemical equilibrium when its concentration is unequal on the two sides of the membrane.

A membrane potential is a delocalized parameter for any given membrane and acts on all ions distributed across a membrane. It therefore follows that a membrane potential generated by the translocation of one ion will affect the electrochemical equilibrium of all ions distributed across the membrane. The membrane potential generated, for example, by proton translocation can therefore be detected by a second ion. If the second ion only permeates by a simple electrical uniport, it will redistribute until its electrochemical equilibrium is regained, and the resulting ion distribution will enable the membrane potential to be estimated from Eq. 3.36. The membrane potential will not be appreciably perturbed by the distribution of the second ion provided the

Table 3.4   The equilibrium distribution of ions permeable by passive uniport across a membrane as a function of $\Delta\psi$ and the charge carried by the ion

$\Delta\psi$ mV	$[X]_{in}/[X]_{out}$			
			Ion charge	
	-1	0	1	2
30	0.3	1	3	10
60	0.1	1	10	100
90	0.03	1	30	1 000
120	0.01	1	100	10 000
150	0.003	1	300	100 000
180	0.001	1	1000	1 000 000

latter is present at low concentration. This is because there is steady-state proton translocation and any transient drop in membrane potential following redistribution of the second ion is compensated by the proton pumping.

This is the principle for most determinations of $\Delta\psi$ across energy-transducing membranes (see Section 4.2.1). The equilibrium ion distribution varies with $\Delta\psi$ as shown in Table 3.4. Note that:

- Anions are excluded from a negative compartment (e.g. the mitochondrial matrix).
- Cation accumulation is an exponential function of $\Delta\psi$.
- Divalent cations are accumulated to much higher distribution ratios than monovalent cations.

3.7.2 Weak acids, weak bases and ΔpH

An electroneutrally permeant species will be unaffected by $\Delta\psi$ and will come to equilibrium when its concentration gradient is unity (Table 3.4). Weak acids and bases (i.e. those with a pK between 3 and 11) can often permeate in the uncharged form across bilayer regions of the membrane (Chapter 2) while the ionized form remains impermeant, even though it may be present in great excess over the neutral species. As a result the *neutral* species (protonated acid or deprotonated base) equilibrates without regard to $\Delta\psi$. However, if there is a ΔpH, the Henderson–Hasselbalch equation requires that the concentration of the *ionized* species must differ (Fig. 3.5).

In summary, weak acid anions become concentrated in the acidic compartment, while protonated bases concentrate in the alkaline compartment. This principle is widely used to determine ΔpH across energy-transducing membranes (Section 4.2.1).

(a) Weak acids

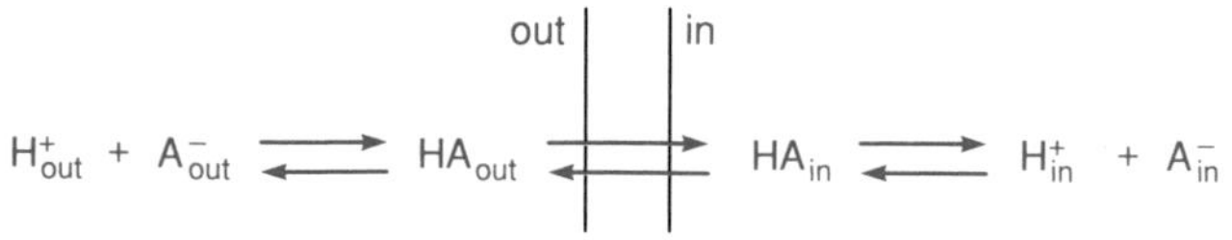

$$\text{out} \quad | \quad \text{in}$$

$$H^+_{out} + A^-_{out} \;\rightleftharpoons\; HA_{out} \quad\rightleftharpoons\quad HA_{in} \;\rightleftharpoons\; H^+_{in} + A^-_{in}$$

If pK of acid is the same in both compartments:

$$K = \frac{[H^+]_{out}\,[A^-]_{out}}{[HA]_{out}} = \frac{[H^+]_{in}\,[A^-]_{in}}{[HA]_{in}}$$

At equilibrium $[HA]_{out} = [HA]_{in}$

$$\therefore\; [H^+]_{in} / [H^+]_{out} = [A^-]_{out} / [A^-]_{in}$$

(b) Weak bases

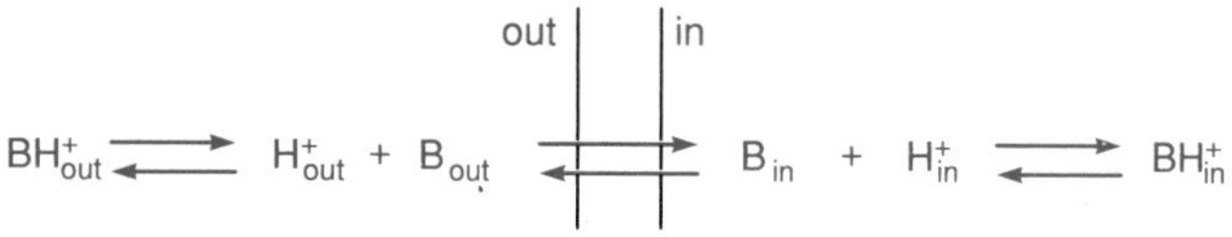

$$\text{out} \quad | \quad \text{in}$$

$$BH^+_{out} \;\rightleftharpoons\; H^+_{out} + B_{out} \quad\rightleftharpoons\quad B_{in} + H^+_{in} \;\rightleftharpoons\; BH^+_{in}$$

$$K = \frac{[H^+]_{out}\,[B]_{out}}{[BH^+]_{out}} = \frac{[H^+]_{in}\,[B]_{in}}{[BH^+]_{in}}$$

At equilibrium $[B]_{out} = [B]_{in}$

$$\therefore\; [H^+]_{in} / [H^+]_{out} = [BH^+]_{in} / [BH^+]_{out}$$

Figure 3.5 The equilibrium distribution of electroneutrally permeant weak acids and bases as a function of ΔpH.
(a) Weak acid distribution: the concentration of the protonated HA is the same on both sides of the membrane, while the anion is concentrated in the alkaline compartment. (b) Weak base distribution: the concentration of the deprotonated B is the same on both sides of the membrane, while the protonated cation is concentrated in the acidic compartment.

3.8 DIFFUSION POTENTIALS, DONNAN POTENTIALS AND SURFACE POTENTIALS

There are two ways in which a true, bulk-phase membrane potential (i.e. trans-membrane electrical potential difference) may be generated. The first is by the operation of an electrogenic ion pump such as operates in energy-transducing membranes. The second is by the addition to one side of a membrane of a salt, the cation and anion of which have unequal permeabilities. The more permeant species will tend to diffuse through the membrane ahead of the counter-ion and thus create a *diffusion potential*. Diffusion potentials may be created across energy-transducing membranes, for example by the addition of external KCl in the presence of valinomycin which provides permeability for K^+, thus generating a $\Delta\psi$, positive inside.

3.8.1 Eukaryotic plasma membrane potentials

In energy-transducing organelles, diffusion potentials tend to be transient due to the rapid movement of counter-ions and are not in general physiologically significant, in contrast to eukaryotic plasma membranes where the generally slow transport processes enable potentials to be sustained for several hours. In this case the diffusion potentials due to the maintained concentration gradients across the plasma membrane play the dominant role in determining the membrane potential. In the case where K^+, Na^+ and Cl^- gradients exist across the membrane, the membrane potential is a function of the ion gradients weighted by their permeabilities, and is given by the Goldman equation:

$$\Delta\psi = \frac{2.3RT}{F} \log_{10} \frac{(P_{Na}[Na^+]_{out} + P_K[K^+]_{out} + P_{Cl}[Cl^-]_{in})}{(P_{Na}[Na^+]_{in} + P_K[K^+]_{in} + P_{Cl}[Cl^-]_{out})} \qquad [3.37]$$

Note that if only a single ion is permeant, this equation reduces to the Nernst equation (Eq. 3.36).

3.8.2 Donnan potentials

The limiting case of a diffusion potential occurs when the counter-ion is completely impermeant. This condition pertains in mitochondria due to the 'fixed'

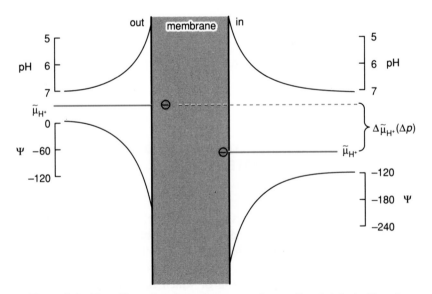

Figure 3.6   The effect of surface charges on the profile of $\Delta\psi$, ΔpH and Δp in the vicinity of the membrane.
Note that although $\Delta\psi$ and ΔpH both change close to the membrane, Δp ($\equiv \Delta\tilde{\mu}_{H^+}$) is unaffected by surface charge. This means that a proton close to the surface of the membrane is in electrochemical equilibrium with the bulk phases.

negative charges of the internal proteins and phospholipids. As a result, when the organelles are suspended in a medium of low ionic strength, such as sucrose, and an ionophore such as valinomycin is added, the more mobile cations attempt to leave the organelle until the induced potential balances the cation concentration gradient. This is a stable Donnan potential.

3.8.3 Surface potentials

Surface potentials are quite distinct from the above. Due to the presence of fixed negative charges on the surfaces of energy-transducing membranes, the proton concentration in the immediate vicinity of the membrane is higher than in the bulk phase (Fig. 3.6). However, Δp is not affected, since the increased proton concentration is balanced by a decreased electrical potential. The proton electrochemical potential difference across the membrane, Δp, is thus unaffected by the presence of surface potentials, although membrane-bound indicators of $\Delta \psi$, such as the carotenoids of photosynthetic membranes (Section 6.3), might be influenced.

The proton circuit of brown adipose tissue (BAT) mitochondria has a leak which can be plugged by nucleotides such as GDP. This is part of the thermogenic mechanism enabling fatty acids to be oxidized to acetate for heat. The Cl^- permeability is puzzling

4 THE CHEMIOSMOTIC PROTON CIRCUIT

4.1 INTRODUCTION

The circuit of protons linking the primary generators of protonmotive force with the ATP synthase was introduced in Chapter 1. The purpose of the present chapter is to discuss the functioning of the proton circuit in a wide range of chemiosmotic energy transductions. The close analogy between the proton circuit and the equivalent electrical circuit (Fig. 1.3) will be emphasized, not only as a simple model but also because the same laws govern the flow of energy around both circuits.

In an electrical circuit the two fundamental parameters are potential difference (in volts) and current (in amps). From measurements of these functions other factors may be derived, such as the rate of energy transmission (in watts) or the resistance of components in the circuit (in ohms). In Fig. 4.1 a simple electrical circuit is shown, together with the analogous proton circuit across the mitochondrial inner membrane (the circuit operating across a photosynthetic or bacterial membrane would be very similar).

In an open circuit (Fig. 4.1a), electrical potential is maximal, but no current flows as the redox potential difference generated by the battery is precisely balanced by the back-pressure of the electrical potential. The tight coupling of the redox reactions within the battery to electron flow prevents any net chemical reaction. In the case of the mitochondrion, the proton circuit is open-circuited when there is no pathway for the protons extruded by the respiratory chain to re-enter the matrix (for example when the ATP synthase is inhibited or when there is no turnover of ATP). As with the electrical circuit

MITOCHONDRION   ELECTRICAL ANALOGUE

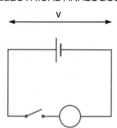

(a) Open circuit. Current zero (no respiration). Potential (Δp) is maximal.

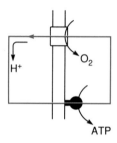

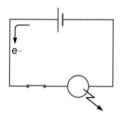

(b) Circuits completed, current flows (respiration occurs). Useful work is done
(ATP is synthesized). Potential (Δp) is less than maximal.

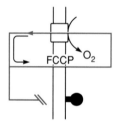

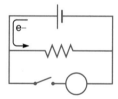

(c) Short-circuit introduced. Energy is dissipated, potential is low, current (respiration) is high.

**Figure 4.1   The regulation of the mitochondrial proton circuit by analogy
to an electrical circuit.**
(a) Open circuit, zero current (no respiration), potential (Δp) maximal. (b)
Circuits completed, current flows (respiration occurs), useful work is done
(ATP is synthesized). Potential (Δp) decreases slightly. (c) Short-circuit
introduced, energy dissipated, potentials are low, current (respiration) is
high.

the potential across the membrane is maximal under these conditions, and
there would be a thermodynamic equilibrium between the proton-translocating
regions of the respiratory chain (Section 3.6.2) and Δp (allowing for the
H^+/e^- stoichiometry of the interconversion; (Section 3.6.2). As the redox
reactions are tightly coupled to proton extrusion there would be no respiration
in this condition.

In Fig. 4.1b the electrical and proton circuits are shown operating normally
and performing useful work. The potential is slightly less than under open-

circuit conditions, as the net driving force which enables the battery or respiratory chain to operate is the slight disequilibrium between the redox potential difference available and the potential in the circuit. The 'internal resistance' of the battery may be calculated from the drop in potential required to sustain a given current. Analogously, the 'internal resistance' of the respiratory chain may be estimated, and is found to be very low (see Fig. 4.14).

An electrical circuit may be shorted by introducing an additional low-resistance pathway in parallel with the existing circuit (Fig. 4.1c). Current can now flow from the battery without having to do useful work, the energy being dissipated as heat. This 'uncoupling' can be accomplished in the proton circuit by the addition of proton translocators (Section 2.3.5a), enabling respiration to occur without stoichiometric ATP synthesis, while a specialized class of mitochondria, in brown adipose tissue, possess a unique proton conductance pathway performing an analogous function (Section 4.7.1).

It is an oversimplification to consider the mitochondrial respiratory chain as a single generator of Δp because the chain consists of three proton pumps operating in parallel from the standpoint of the proton circuit and in series with respect to the electron flow (Fig. 4.2). These pumps, which will be described in more detail in Chapter 5, are known as complexes I, III and IV. It should be noted that it is possible to introduce (e.g. with ascorbate plus TMPD) or remove electrons at the interfaces of an individual proton pump, allowing it to be studied in isolation.

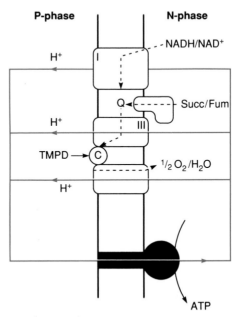

Figure 4.2 The mitochondrial respiratory chain consists of three proton pumps (complexes I, III and IV) which act in parallel with respect to the proton circuit and in series with respect to the electron flow.
Solid blue lines: pathway of proton flux. Dotted line: pathway of electron transfer. For redox couples see Table 3.2.

4.2 THE MEASUREMENT OF PROTONMOTIVE FORCE

Quantification of Δp has been important for the chemiosmotic theory, since a single demonstration of net ATP synthesis in the absence of a sufficient Δp would have been sufficient to demolish the entire edifice. Furthermore, it has been the observation of apparent inadequacies in the magnitude of the Δp which has fuelled the debate on 'localized chemiosmosis' (Section 4.9).

All techniques for the determination of Δp involve the separate estimation of $\Delta\psi$ and ΔpH (Section 3.7). $\Delta\psi$ can be calculated by direct determination of the concentration gradient at equilibrium of an ion that permeates by electrical uniport (Fig. 2.1) by application of the Nernst equation (Eq. 3.36). Alternatively a diffusion potential (Section 3.8) generated by an ion gradient, such as K^+ in the presence of valinomycin, can calibrate a spectroscopic or fluorimetric indicator of $\Delta\psi$. ΔpH is generally calculated from the equilibrium distribution of electroneutrally permeant weak acids and bases (Section 3.7.2) which may be either radiolabelled or fluorescent.

Considerable care must be taken in the selection of appropriate indicators. Firstly, to measure $\Delta\psi$ the ion should be of the correct charge (Table 3.4) to be accumulated: a cation if the interior is negative (mitochondria or bacterial cells) or an anion if the interior is positive with respect to the medium (e.g. SMPs or chromatophores). Secondly, it must be possible to calculate the free indicator concentration within the organelle, by choosing either an indicator which is not bound or one whose activity coefficient may be readily calculated. Thirdly, the indicator must readily achieve electrochemical equilibrium and not be capable of being transported by more than one mechanism. Fourthly, the indicator should disturb the gradients as little as possible. Finally, the indicator should not be metabolized.

For the measurement of ΔpH, the above conditions should also be satisfied, with the exception that the indicator should be a weak acid to be accumulated within organelles with an alkaline interior, and a weak base to be accumulated in an acidic compartment (Section 3.7.2).

Once an equilibrium distribution of an indicator has been achieved it must be measured. This can be accomplished by rapid separation of the organelle from the incubation medium, by continuously monitoring the fall in indicator concentration in the incubation as the ion is accumulated, or by making use of the altered spectral properties of the indicator when accumulated by an energy-transducing membrane. Examples of these techniques will now be considered.

4.2.1 The determination of Δp by ion-specific electrodes and radio-isotopes

Bioenergetic organelles and bacterial cells are too small to be impaled by microelectrodes, and ion gradients must be calculated from the fall in external concentration (or the rise in the internal concentration) of an indicator ion as it is accumulated. This means that both the internal volume and the activity coefficient of the ion must be known.

The first determination of Δp in mitochondria (Mitchell and Moyle, 1969) employed pH- and K^+-specific electrodes in an anaerobic, low-K^+ incubation. Valinomycin was present to create a high electrical uniport for K^+, and $\Delta\psi$ was calculated from the K^+ uptake occurring on the addition of an extended pulse of O_2. ΔpH was estimated from the parallel proton-extrusion. A value of 228 mV was obtained for Δp for mitochondria respiring under 'open-circuit' conditions in the absence of ATP synthesis (state 4, see Section 4.6), and a value in this range has stood the test of time ever since.

The technique can be modified for radioactive assay by substituting the β-emitter ^{86}Rb for K^+ and by using labelled weak acids (such as acetate) and bases (e.g. methylamine) to determine ΔpH. After silicone oil centrifugation to separate the mitochondria from the medium (Fig. 4.3) the concentration gradient of the isotopes across the inner membrane can be calculated. In these experiments an additional indicator, such as $[^{14}C]$ sucrose, must be included in the incubation medium. Sucrose is not permeable and serves to indicate the volume of contaminating medium which is pelleted together with the mitochondria. It is also necessary to estimate the matrix volume. This is difficult to do with precision although comparison of the sucrose-permeable and ^3H_2O-permeable spaces gives a reasonably accurate measure (Fig. 4.3). A

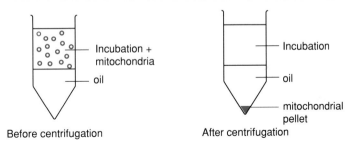

Before centrifugation   After centrifugation

Figure 4.3 Silicone-oil centrifugation of mitochondria for the determination of $\Delta\psi$.
Mitochondria are incubated under the desired conditions in a K^+-free medium containing valinomycin, $^{86}Rb^+$, $[^{14}C]$ sucrose and ^3H_2O. An aliquot of the incubation is added to an Eppendorf tube containing silicone oil and centrifuged for about 1 min at 10 000 g. The mitochondria form a pellet under the oil and can be solubilized for liquid scintillation counting. An aliquot of the supernatant is also counted. The $[^{14}C]$ sucrose in the incubation allows the sucrose-permeable space, V_s, in the pellet to be calculated. This gives the extra-matrix contamination of the pellet with incubation medium (i.e. extra-mitochondrial space plus intermembrane space). The difference between the ^3H_2O-permeable space in the pellet (V_h) and V_s gives the sucrose-*im*permeable space, which is taken to represent the matrix volume (as water but not sucrose can permeate the inner membrane). If the apparent ^{86}Rb space in the pellet (V_{Rb}) is calculated similarly, then $\Delta\psi$ can be calculated from the Nernst equation:

$$\Delta\psi = 2.3RT \log_{10} \frac{(V_{Rb} - V_s)}{(V_h - V_s)}$$

major error, however, lies in estimating the activity of the cations within the mitochondrial matrix. The matrix is enormously concentrated: about 0.5 mg of protein is concentrated into 1 μl, i.e. 500 g/l! In such a concentrated gel it is inevitable that the ions behave non-ideally.

The use of valinomycin in the above experiments has the disadvantage that $\Delta\psi$ is artifactually clamped at a value corresponding to the Nernst equilibrium for the pre-existing K^+ gradient across the membrane. This can be avoided by the use of phosphonium cations, such as TPP^+ (tetraphenyl-phosphonium) (Fig. 4.4). The positive charge is sufficiently delocalized by the π-orbital system and screened with hydrophobic groups to enable the cation to permeate bilayer regions (Section 2.3.6). A representative lipophilic and permeant anion (TPB^-) is also shown in Fig. 4.4, whilst other suitable anions are SCN^- and ClO_4^- which tend to exhibit less non-specific binding than TPB^-.

TPP^+ accumulation can be measured isotopically, or alternatively a TPP^+-selective electrode can be constructed allowing a continuous monitoring of the external concentration from which the internal concentration and hence $\Delta\psi$ can be calculated via the Nernst equation (Eq. 3.36) (Fig. 4.5). The activity coefficients of these indicators deviate widely from unity when accumulated in mitochondria or bacteria, and it is necessary to calibrate the response, e.g. by quantifying uptake in the presence of valinomycin and known K^+ gradients. Often the pH component of mitochondrial Δp is minimized by the inclusion of high Pi concentrations; the approximation is then made that $\Delta\psi$ accounts for the entire Δp.

As an aside we note that 1-Methyl-4-phenylpyridinium ion (MPP^+) is lipophilic cation which is formed *in vivo* in certain brain areas from 1-methyl-4-phenyl-1,2,3,6-tetrahydropyridine (MPTP) and which induces a Parkinson's-like disease by destroying dopamine-secreting nerves. MPP^+ acts like TPP^+ and, particularly in the presence of a counter-ion such as tetraphenylboron, accumulates within the mitochondria, collapsing $\Delta\psi$ and also directly inhibiting complex I (Ramsay *et al.*, 1986).

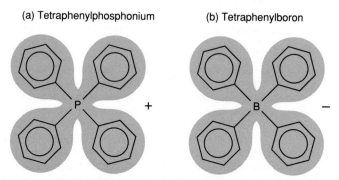

Figure 4.4   **Representative example of a lipophilic cation and anion.** Tetraphenylphosphonium (TPP^+) (a) and tetraphenylboron (TPB^-) (b) are lipophilic ions which can cross hydrophobic membranes despite their charge. The shaded area represents the extent of the π-orbital systems.

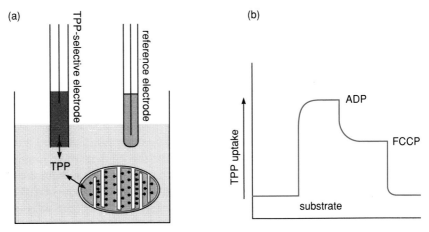

Figure 4.5   Tetraphenylphosphonium and the determination of $\Delta\psi$.
(a) A PVC membrane may be made selectively permeable to TPP^+ by incorporating tetraphenylboron during its preparation. This can now be assembled into a TPP^+ electrode which can measure the concentration of TPP^+ in an incubation. As the mitochondria accumulate the cation in response to their membrane potential the electrode will detect the fall in external $[TPP^+]$. (b) Typical trace showing the generation of a membrane potential when substrate is added to mitochondria. Addition of ADP causes a slight fall in $\Delta\psi$, which is collapsed by the subsequent addition of the protonophore FCCP.

4.2.2   Optical indicators of $\Delta\psi$ and ΔpH

(a) Intrinsic indicators of $\Delta\psi$

A $\Delta\psi$ of some 200 mV corresponds to an electrical field across the energy-transducing membrane in excess of 300 000 V cm^{-1}. It is not surprising, therefore, that certain natural membrane constituents respond to the electrical field by altering their spectral properties. Such electrochromism is due to the effect of the imposed field on the energy levels of the electrons in a molecule. The most widely studied of these intrinsic probes of $\Delta\psi$ are the carotenoids of photosynthetic energy-transducing membranes (Jackson, 1988). Carotenoids are a heterogeneous class of long-chain, predominantly aliphatic pigments which are found in both chloroplasts and photosynthetic bacteria. Their roles include light-harvesting (Chapter 6) and protection against oxidative damage of the photosynthetic apparatus; they can trap reactive excited states of the oxygen molecules. A common feature of carotenoid molecules is a central hydrophobic region with conjugated double bonds which allows delocalization of electrons and gives carotenoids a characteristic visible spectrum. The shifts in their absorption spectra in response to the membrane potentials experienced by energy-transducing membranes are only a few nanometres and so the signal is usually detected by dual wavelength spectroscopy (see Fig. 5.2). The carotenoids respond with extreme rapidity (ns or less) and enable the primary electrogenic events to be followed (Section 6.2). The carotenoid band shift can be calibrated, especially with chromatophores, by imposing K^+ diffusion potentials in the dark by adding KCl and valinomycin. This gives a potential, positive inside, that will vary with the KCl concentration according to the Nernst equation (Eq. 3.36). Illumination causes translocation of protons into the lumen of such chromatophores (Chapter 1) and thus a shift in the same direction as caused by imposed K^+ diffusion potentials. Thus the calibration allows one to quantitate the light-dependent shift in terms of a membrane potential. One limitation of the carotenoids, however, is that, being bound in the membrane, they only detect the field in their immediate environment, which need not correspond to the bulk-phase membrane potential difference measured by distribution techniques, particularly as surface potential effects (Section 3.8.3) could be significant. In this context it is generally found that in chromatophore membranes the carotenoid shift gives much larger values of $\Delta\psi$ than ion distribution methods. However, there is no decisive evidence that surface potential effects account for this important and unresolved discrepancy.

(b) Extrinsic indicators of $\Delta\psi$ and ΔpH

Continuous, non-destructive optical techniques rely on the spectral changes which occur when certain charged, lipophilic dyes accumulate in response to the membrane potential. This enables the extent of association to be followed in incubations without the necessity to separate the organelles. These changes are spectral shifts due to concentration-dependent changes in the extent of dye aggregation within the cell. Since they are large planar molecules (Fig. 4.6), the probes often possess the ability to form stacks of molecules when in locally high concentration. This stacking reduces their ability to absorb light and underlies at least some of the observed spectral changes. The complex nature of the probe response means that many factors other than $\Delta\psi$ or ΔpH can interfere, and great care must therefore be taken to control the conditions under which they are used. The spectral response must always be calibrated, for example by reference to a known K^+ diffusion potential (Section 3.8) as described for carotenoid calibration (Section 4.2.2a). No extrinsic probe has yet been described which acts like the intrinsic carotenoids, i.e. remaining fixed in the membrane and altering its spectrum in response to the applied field.

The cationic carbocyanines (Fig. 4.6) respond very rapidly to changes in membrane potential, while the anionic bisoxonols (Fig. 4.6) are very sensitive, with a large fluorescent yield. They respond, however, only slowly to changes in membrane potential, requiring up to 1 min to re-equilibrate after a transient.

ΔpH can be estimated from the fluoresence quenching of certain acridine dyes (Fig. 4.6) which are weak bases and so will tend to accumulate on the acidic side of a membrane where, for reasons that are not fully understood, their fluorescence is quenched. There are often problems of quantitating such quenching in terms of a pH gradient but they can be useful qualitative

(a) Cationic
   1. Phenosafranine

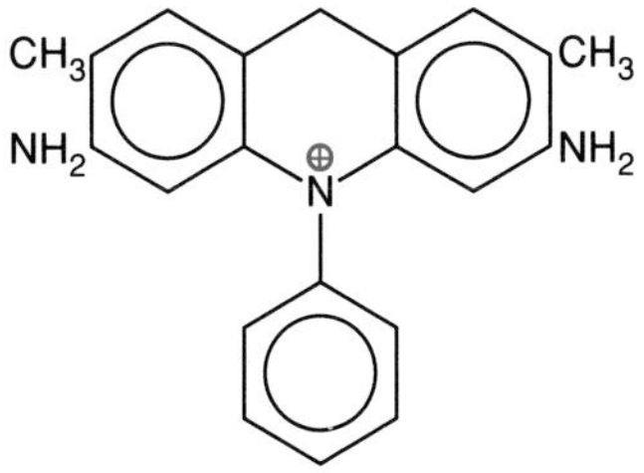

2. Cyanine dyes, e.g.

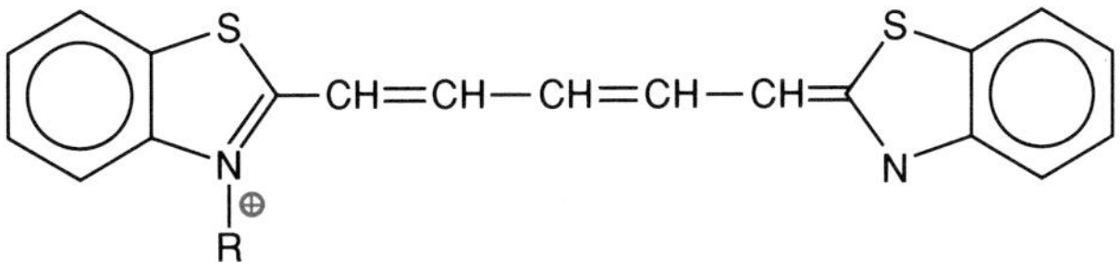

(b) Anionic
   1. Oxonols, e.g.

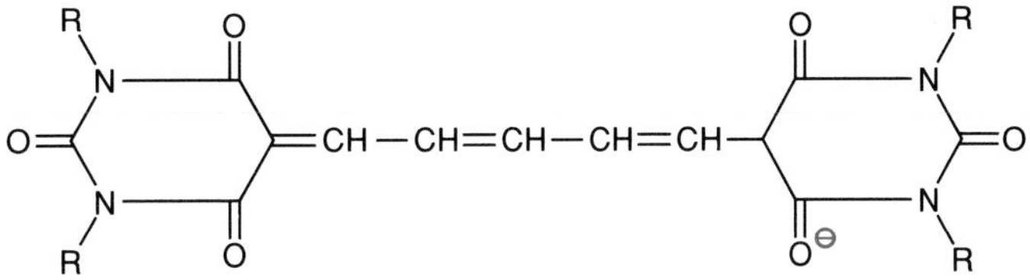

(c) Weak bases
   1. 9-aminoacridine

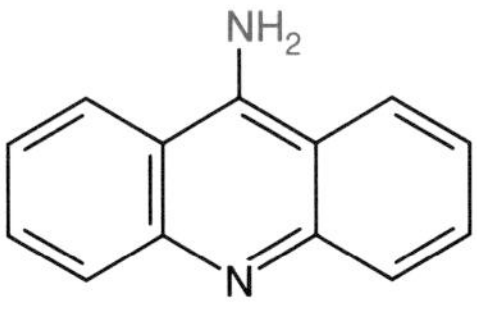

Figure 4.6   Optical indicators of $\Delta\psi$ and ΔpH.
(a) Cationic dyes; (b) anionic dyes; (c) weak base dyes.

probes because small amounts of membranes may be assayed without any requirement to separate the membranes from the suspending medium.

4.2.3 ΔpH determination by ^{31}P-NMR

Nuclear magnetic resonance (NMR) can be used to obtain a direct measurement of the pH inside and outside a cell or organelle, and is free of some of the drawbacks inherent in the more invasive use of weak acids or bases. The basis of the technique is that the resonance energy of the phosphorus nucleus in Pi varies according to the protonation state of the latter. As the pK_a for $H_2PO_4^-/HPO_4^{2-}$ is 6.8 the technique can report pH values in the range 6–7.5. The NMR signal is the average for the two ionization states, since proton exchange is fast on the NMR timescale. If there is phosphate in both the external and internal phases a ΔpH can be calculated. The drawback is that NMR is an insensitive method and millimolar concentrations of Pi and thus thick cell suspensions are required, with attendant problems of supplying oxygen and substrates.

4.2.4 What decides the contribution of Δψ and ΔpH to Δp?

The events which regulate the partition of Δp between $\Delta \psi$ and ΔpH are summarized in Fig. 4.7. The mitochondrion will be used as an example, but the discussion is equally valid for other energy-transducing systems.

Starting from a 'de-energized' state of zero protonmotive force, the operation of a proton pump in isolation leads to the establishment of a Δp in which

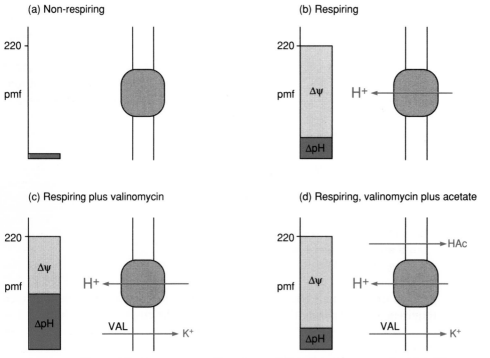

Figure 4.7   Factors controlling the partition of Δp between $\Delta \psi$ and ΔpH. Starting from a de-energized mitochondrion (a), the initiation of respiration (b) leads to a high $\Delta \psi$ and low ΔpH, since the electrical capacitance of the membrane is very low. In a high-$[K^+]$ medium, valinomycin collapses $\Delta \psi$ (c) and allows a high ΔpH to build up. If a permeant weak acid is additionally present (d) ΔpH will collapse and extensive swelling may occur.

the dominating component is $\Delta\psi$ (Fig. 4.7a). The electrical capacity of the membrane is such that the net transfer of 1 nmol H^+ (mg protein)$^{-1}$ across the membrane establishes a $\Delta\psi$ of about 200 mV. The pH buffering capacity of the matrix is about 20 nmol of H^+ (mg protein)$^{-1}$ per pH unit, and the loss of 1 nmol H^+ will only increase the matrix pH by 0.05 units (i.e. equivalent to 3 mV). Δp will thus be about 99% in the form of a membrane potential (Fig. 4.7b). In the absence of a significant flow of other ions it will stay this way in the steady state.

If an electrically permeant ion such as Ca^{2+}, or K^+ plus valinomycin, is now added (Fig. 4.7c), its accumulation in response to the high $\Delta\psi$ will tend to dissipate the membrane potential and hence lower Δp. The respiratory chain responds to the lowered Δp by a further net extrusion of protons, thus restoring Δp. Because of the pH buffering capacity of the matrix, the uptake of 20 nmol K^+ (or 10 nmol Ca^{2+}), balanced by the extrusion of 20 nmol H^+, will lead to the establishment of a ΔpH of about -1 unit (equivalent to 60 mV). As the respiratory chain can only achieve the same total Δp as before, this means that the final $\Delta\psi$ must be nearly 60 mV lower than before uptake of the cation. Thus cation uptake leads to a redistribution from $\Delta\psi$ to ΔpH. The lowered $\Delta\psi$ means that cation uptake under these conditions becomes self-limiting, as the driving force steadily decreases until equilibrium is attained. For example, the uptake of Ca^{2+} by mitochondria in exchange for extruded protons is limited to about 20 nmol (mg protein)$^{-1}$ (Section 8.3), by which the time $\Delta\psi$ has decreased (and $-60\,\Delta$pH increased) by about 120 mV. In contrast, we note that when TPP^+ is used at low concentrations as a probe for $\Delta\psi$ the uptake of the cation is sufficiently small to avoid perturbation to Δp and ΔpH (see Section 3.7.1). Higher concentrations of TPP^+ would cause the steady-state Δp to decline with a compensatory increase in ΔpH.

The third event which can influence the relative contributions of $\Delta\psi$ and ΔpH is the redistribution of electroneutrally permeant weak acids and bases (Fig. 4.7d). Uptake of a weak acid in response to the ΔpH created by prior cation accumulation dissipates the pH gradient and allows the respiratory chain to restore $\Delta\psi$. However, the situation can now be reached where both cation and anion have been accumulated, and if amounts are large, osmotic swelling of, for example, the mitochondrial matrix (Section 2.8) results. This does not occur when the cation and anion are respectively Ca^{2+} and Pi, as formation of a non-osmotically active calcium phosphate complex prevents an increase in internal osmotic pressure (Section 8.3).

It is clear from the above discussion that $\Delta\psi$ and pH indicators themselves, being ions, weak acids or weak bases, can disturb the very gradients to be measured unless care is taken. This is particularly true in the presence of valinomycin, as the ionophore brings into play the high endogenous K^+ of the matrix, with the result that $\Delta\psi$ will become clamped at the value given by the initial K^+ gradient. This risk is less apparent with cations such as tetraphenylphosphonium (TPP^+), which can be employed at very low concentrations.

4.2.5 The magnitude of the protonmotive force and assessment of its thermodynamic competence

Table 4.1 presents some approximate values for the protonmotive force as well as its two components, $\Delta\psi$ and ΔpH, that have been determined for a number of energy-transducing membranes. Also shown are values for ΔG_p and H^+/ATP obtained by comparison of the two parameters. Although some experimental determinations of Δp deviate (Section 4.9) from the pattern shown in Table 4.1, the general conclusion is that the Δp is thermodynamically competent to support ATP synthesis with the H^+/ATP values shown. The latter values are broadly consistent with those obtained by other methods.

The same conclusion can be drawn if the protonmotive force is compared with the uptake of a solute, e.g. lactose in *E. coli*. A point to note is that SMPs appear to sustain a lower ΔG_p than mitochondria. This is consistent with the

Table 4.1   Values of Δp and ΔG_p together with calculated H^+/ATP ratios for energy transducing membranes.

Membrane	$\Delta\psi$ (mV)	ΔpH (units)	Δp (mV)	ΔG_p (kJ mol^{-1})	H^+/ATP
Liver mitochondria	170[a]	⩽0.5[b]	⩽200	66	⩾3.4[*]
Submitochondrial particles (heart)	150[c]	⩽0.5[d]	⩽180	45	⩾3.1
Thylakoids	0[c]	3.3[d]	195	60	3.1
Inside-out vesicles from *P. denitrificans*	160[c]	⩽0.5[d]	⩽190	60	⩾3.2
E. coli cells at pH 7.5	140[a]	⩽0.5[b]	⩽170	40	≠

These are consensus values estimated from those in the literature. As described in this chapter the experimental determinations of the components of Δp are indirect and subject to error. Thus it cannot be discounted that literature values that are either higher or lower than those quoted here may ultimately prove to be better estimates. In all cases the steady state Δp was generated by electron transport through provision of a suitable respiratory substrate except for thylakoids which were illuminated. $\Delta\psi$ is negligible under steady state conditions for thylakoids because of chloride and other ion movements (see Chapters 3 and 6).
[a] From distribution of TPP$^+$ or ^{86}Rb$^+$; [b] from distribution of acetate or 5,5,dimethyl-oxazolidene 2,4 dione; [c] from distribution of SCN$^-$ or ClO$_4^-$; [d] from alkylamine distribution; ≠, cannot be meaningfully calculated because in the intact cells ΔG_p will not approach its equilibrium value with Δp owing to continual consumption of ATP; [*] this value will include the H^+/ATP ratio for the ATP synthase plus the net movement of one proton associated with adenine nucleotide and phosphate translocation (see Sections 3.6.3 and 8.4.5 and Fig. 8.5). If ΔG_p is an underestimated and/or Δp overestimated the value of 3.4 will also be an underestimate for a value of 4.

In the experiments quoted an electron transport-dependent Δp was compared with the ΔG_p generated by the energy transducing membranes. If the ΔG_p had not reached its maximum value that would be at equilibrium with Δp then the values of H^+/ATP would tend to be underestimates. However, with submitochondrial particles comparison of ΔG_p, held constant by an ATP regenerating system, with Δp that was generated by ATP hydrolysis, a procedure that would tend to give an overestimate of H^+/ATP because Δp would tend to be below its equilibrium value, also indicated an H^+/ATP ratio of the order of 3, thus strengthening the case for accepting this stoichiometry (see also Chapter 7).

contribution of the adenine nucleotide translocator and phosphate transport in extra-mitochondrial ATP synthesis (Sections 3.6.3 and 8.4.5).

4.3 THE STOICHIOMETRY OF PROTON EXTRUSION BY THE RESPIRATORY CHAIN

The proton current generated by the respiratory chain cannot be determined directly under steady-state conditions as there is no means of detecting the flux of protons when the rate of proton efflux exactly balances that of re-entry. It is, however, possible to measure the initial ejection of protons which accompanies the onset of respiration before proton re-entry has become established. By making a small and precise addition of O_2 to an anaerobic mitochondrial suspension in the presence of substrate and monitoring the extent of proton extrusion with a rapidly responding pH electrode, one can thus obtain a value for the H^+/O stoichiometry for the segment of the respiratory chain between the substrate and O_2.

4.4 EXPERIMENTAL DETERMINATION OF H^+/O RATIOS

The practical details of an experiment to determine H^+/O are shown in Fig. 4.8. A number of precautions are necessary: first, an electrical cation permeability has to exist to allow charge-compensation of the proton extrusion, which would otherwise be limited by the rapid build-up of $\Delta\psi$. Secondly, the pulse of O_2 must be small enough to prevent ΔpH from saturating. Thirdly, however rapid the pulse of respiration, some protons will leak back across the membrane (and thus be undetected) before the burst of respiration is completed. These protons must be allowed for: the problem is enhanced if Pi (which is nearly always present in mitochondrial preparations) is allowed to re-enter the mitochondrion during the O_2 pulse. $H^+:Pi^-$ symport (remember this is the same as OH^-/Pi^- antiport) is extremely active in most mitochondria, and the ΔpH-induced Pi uptake results in movement of protons into the matrix and hence an underestimate of H^+/O stoichiometry. Inhibition of the phosphate symport by N-ethylmaleimide significantly increased the observed H^+/O ratio.

Electron acceptors other than O_2 can be used to select limited regions of the mitochondrial respiratory chain, allowing a $H^+/2e^-$ ratio for the span to be obtained. Additionally, the charge stoichiometry (q^+/O or $q^+/2e^-$) may be determined instead of the proton stoichiometry by quantifying the compensatory movement of K^+ in the presence of valinomycin. Charge and proton stoichiometry are not necessarily synonymous, since electrons can enter and leave respiratory complexes on opposite sides of the membrane. For example, as will be described in Chapter 5, two electrons enter mitochondrial complex III from the matrix and are delivered to cytochrome *c* on the outer face of the

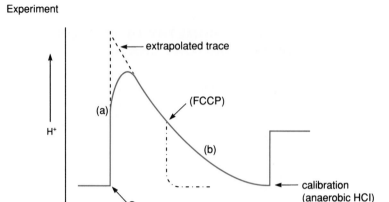

Experiment

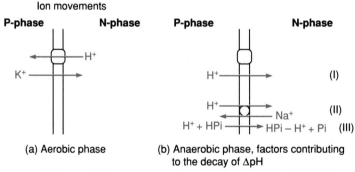

Ion movements

(a) Aerobic phase

(b) Anaerobic phase, factors contributing
to the decay of ΔpH

Figure 4.8   Determination of mitochondrial H⁻/O ratios by the 'oxygen pulse' technique.

A concentrated mitochondrial suspension is incubated anaerobically in a lightly buffered medium containing substrate, valinomycin and a *high* concentration of KCl. The pH of the incubation is continuously monitored by a fast-responding pH electrode. To start the transient, a *small* aliquot of air-saturated medium, containing about 5 nmol O (mg protein)$^{-1}$, is rapidly injected. There is a rapid acidification of the medium as the respiratory chain functions for 2–3 s while using up the added O_2. Valinomycin and K$^+$ are necessary to discharge any Δψ which would limit proton extrusion. When O_2 is exhausted the pH transient decays as protons leak back into the matrix. This decay can be due to (i) proton permeability of the membrane: note that FCCP accelerates the decay, (ii) the action of the endogenous Na$^+$/H$^+$ antiport, (iii) electroneutral Pi entry. The trace must then be corrected by extrapolation to allow for proton re-entry which occurred before the oxygen was exhausted.

membrane; at the same time four protons are pumped from the matrix. The $q^+/2e^-$ ratio is thus 1 while the $H^+/2e^-$ ratio is 2.

An alternative method for determining H^+/O ratios is based on the measurement of the initial rates of respiration and proton extrusion when substrate is added to the substrate-depleted mitochondria. This approach tends to give higher values than the O_2-pulse procedure.

Any stoichiometry determined by the above methods has to satisfy the constraints imposed by thermodynamics. In other words the energy conserved in the proton electrochemical potential has to lie within the limits imposed by the redox span of the proton-translocating region. In addition, the proton-translocating regions of complexes I and III (which will be described in Chapter 5) are known to be in near-equilibrium as they can be readily reversed. Therefore an approximate stoichiometry for these regions can be deduced on purely thermodynamic grounds, knowing ΔE_h (Section 3.8) and the components of Δp. This is a problem associated with the very high H^+/O stoichiometries reported by Lehninger (Lehninger et al., 1979): such values of up to 12 for electrons originating from NADH cannot be reconciled with available data for Δp and ΔE_h.

The determination of $H^+/2e^-$ stoichiometries for the proton-translocating regions of the respiratory chain was for a long time contentious. The direct group-translocation mechanism originally proposed by Mitchell (Fig. 4.9), in which protons are extruded as a consequence of the transfer of electrons from $(H^+ + e^-)$ carriers to e^- carriers, required that the observed stoichiometry did not exceed $2H^+/2e^-$ for each 'loop' (Fig. 4.9). This direct mechanism also makes a number of structural demands which are discussed later. Early measurements of H^+/O ratios by the O_2-pulse method (Mitchell and Moyle, 1967b) gave values of NADH $\rightarrow O_2$ and succinate $\rightarrow O_2$ close to those predicted by the direct mechanism (6 and 4 respectively in Fig. 4.9). However, the findings that N-ethylmaleimide inhibition of the H^+/Pi symporter increased the observed stoichiometry for NADH $\rightarrow O_2$ to between 8 and 10, and that for succinate $\rightarrow O_2$ to 6, together with the observation that the terminal complex of the respiratory chain translocated not only charge (e.g. Fig. 4.9a) but also protons (e.g. Fig. 4.9b), are inconsistent with the original loop model.

Accurate stoichiometries have been notoriously difficult to obtain: the thermodynamic approach of equating the Gibbs energy change in reversible regions of the respiratory chain with the magnitude of the Δp under near-equilibrium conditions is beset with problems in the accurate determination of the latter parameter, and also with the possibility that the 'bulk phase' Δp may not represent the correct thermodynamic potential in a possible 'localized proton circuit' (Section 4.9). Non-steady-state determinations of proton extrusion are also subject to controversy: it is difficult to eliminate the movement of compensatory ions masking the pH change, while some of the few protons which appear per respiratory chain complex in these experiments could conceivably originate from pK changes on the protein itself rather than as a result of transmembrane translocation. Whilst essentially all experiments on complex III (from mitochondria or bacteria) give a stoichiometry of $4H^+/2e^-$ (see

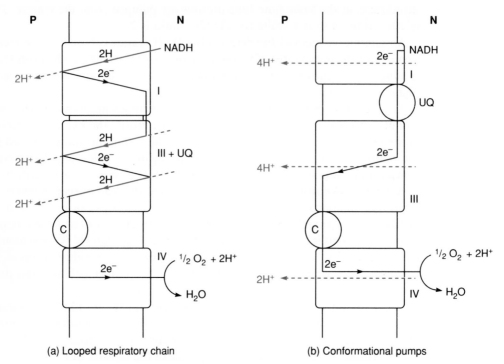

(a) Looped respiratory chain   (b) Conformational pumps

Figure 4.9   Proton and charge stoichiometry in the respiratory chain: the loop versus pump controversy.
In a the classical 'looped' respiratory chain of Mitchell is shown. Note that complex III comprises $1\frac{1}{2}$ loops, while cytochrome oxidase is just the electron-carrying limb of the final loop. Unless an additional loop is proposed within cytochrome oxidase, this mechanism is limited to a H^+/O of 6 for the span $NADH \rightarrow O_2$. In (b) the broad features of a conformational pump respiratory chain are shown. Note that no molecular mechanisms for proton translocation are shown, and that the H^+/O ratio is free of mechanistic, but not thermodynamic, constraints. The ratios shown in (b) are 'consensus' values and may be subject to revision, particularly in the case of complex I.

Section 5.8) there is still some debate about the values for complexes I and IV. Final confirmation of the consensus values in Fig. 4.9b may require the development of a new experimental approach.

4.5 THE STOICHIOMETRY OF PROTON UPTAKE BY THE ATP SYNTHASE

The number of protons translocated by the ATP synthase during the synthesis of one ATP may be calculated either by measuring the transient proton extrusion during the hydrolysis of a small known amount of ATP or by thermodynamic analysis of Δp and ΔG_p for the ATP/ADP + Pi reaction under equilibrium conditions. The transient technique is analogous to the O_2-pulse method (Fig. 4.8) except that ATP is substituted for O_2, but there are a number of additional problems. First, the hydrolysis of ATP in itself generates 'scalar' protons, i.e. on one side of the membrane (Eq. 3.10). Secondly, entry of ATP and exit of ADP and Pi generates 'vectorial' (i.e. translocated) protons (see Section 8.4.5). The second problem may be overcome by working with inverted SMPs (Section 1.4), and the first by adjusting the pH so that there are no scalar protons. Even so, values from 2 to 4 H^+/ATP have been reported.

The thermodynamic approach (Fig. 4.10) has been used for both mitochondria, SMPs, bacterial vesicles and chloroplasts. In the case of intact mitochondria, a value of 2.5 to 4 has been found by this approach. One of the protons can be ascribed to the transport of ADP, Pi and ATP (see above). Note that determination of the H^+/ATP in this way necessitates accepting the validity of the chemiosmotic theory and of the estimates of Δp. Only when there is an independent indication of the magnitude of the H^+/ATP can the comparison of Δp with ΔG_p (Section 4.2.5) be used in support of the chemiosmotic theory.

The stoichiometries for proton extrusion by a segment of the respiratory chain (H^+/$2e^-$) and for proton re-entry during ATP synthesis (H^+/ATP) have

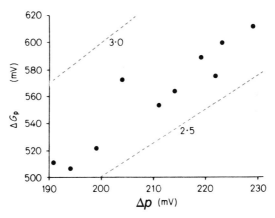

Figure 4.10 Thermodynamic relationship between the extramitochondrial ΔG_p and Δp.
Brown adipose tissue mitochondria were incubated in a medium containing sn-glycerol-3-phosphate as substrate, GDP and albumin to inhibit the proton short-circuit present in these mitochondria (Section 4.7.1) and isotopes to enable Δp to be determined. The equilibrium extra-mitochondrial ATP/ADP ratio was determined isotopically to enable ΔG_p (here expressed in mV) to be calculated. ΔG_p and Δp were compared for a range of low FCCP concentrations. The diagonal dotted lines represent the highest ΔG_p which could be maintained if either 2.5 or 3 protons were used in the synthesis and transport of one ATP. (Data from Nicholls and Bernson, 1977.) If four protons were used (c.f. Fig. 8.5) ΔG_p values should be higher.

to be consistent with the overall observed stoichiometry of ATP synthesis related to electron flow (ATP/2e$^-$):

$$ATP/2e^- = (H^+/2e^-)/(H^+/ATP) \qquad [4.1]$$

or when O_2 is the final acceptor:

$$ATP/O = (H^+/O)/(H^+/ATP) \qquad [4.2]$$

A value for ATP/O of 3 for NADH $\rightarrow O_2$ and 2 for succinate $\rightarrow O_2$ (see Section 4.10), together with a H^+/ATP ratio of 3 to 4 for combined ATP synthesis and export, by mitochondria would require H^+/O ratios of 12 and 8 respectively for NADH and succinate. The relation of currently accepted lower values for H^+/O to the mitochondrial ATP/O ratio is discussed in Section 4.10.2.

4.6 PROTON CURRENT, PROTON CONDUCTANCE AND RESPIRATORY CONTROL

The previous sections have been concerned with the potential term in the proton circuit and with the gearing of the transducing complexes for the generation and utilization of this potential. This section will deal with the factors which regulate the proton current in the circuit.

The current of protons flowing around the proton circuit (J_{H+}) may be readily calculated from the rate of respiration and the H^+/O stoichiometry:

$$J_{H+} = (\delta O/\delta t) \times (H^+/O) \qquad [4.3]$$

For a given substrate, therefore, proton current and respiratory rate vary in parallel, and thus an oxygen electrode (Fig. 4.11) is an effective way of monitoring J_{H+} as long as the H^+/O stoichiometry is known.

The oxygen electrode (Fig. 4.11) has long been the most versatile tool for investigating the mitochondrial proton circuit. Although the electrode only determines directly the rate of a single reaction, the final transfer of electrons to O_2, information on many other mitochondrial processes can be obtained simply by arranging the incubation conditions so that the desired process becomes a significant step in determining the overall rate. Several such steps may be investigated (Fig. 4.12) including:

(a) substrate transport across the membrane
(b) substrate dehydrogenase activity
(c) respiratory chain activity
(d) adenine nucleotide transport across the membrane
(e) ATP synthase activity
(f) proton permeability of the membrane

Three basic states of the proton circuit were shown in Fig. 4.1: open circuit, where there is no evident means of proton re-entry into the matrix; a circuit completed by proton re-entry coupled to ATP synthesis; and a circuit

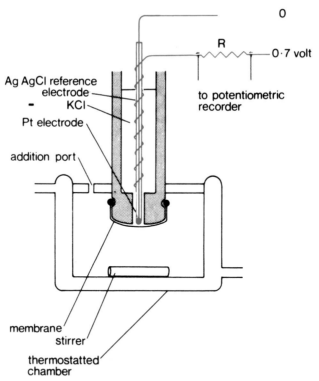

Figure 4.11   The Clark oxygen electrode.
At the Pt electrode, which is maintained 0.7 V negative with respect to the Ag/AgCl reference electrode, O_2 is reduced to H_2O, and a current flows which is proportional to the O_2 concentration in the medium. A thin O_2-permeable membrane prevents the incubation from making direct contact with the electrodes. Since the electrode slowly consumes O_2 the incubation must be continuously stirred to prevent a depletion layer forming at the membrane. The chamber is sealed except for a small addition port. The electrode is calibrated with air-saturated medium and under anoxic conditions following dithionite addition.

completed by a proton leak not coupled to ATP synthesis. These states can readily be created in the oxygen-electrode chamber (Fig. 4.13).

When the oxygen electrode was first being applied to mitochondrial studies, Chance and Williams (1956) proposed a convention following the typical order of addition of agents during an experiment (Fig. 4.13):

State 1: mitochondria alone (in the presence of Pi)
State 2: substrate added, respiration low due to lack of ADP
State 3: a limited amount of ADP added, allowing rapid respiration
State 4: all ADP converted to ATP, respiration slows
State 5: anoxia

The addition of mitochondria to an incubation medium containing Pi (Fig. 4.13) causes little respiration as no substrate is present. Although mitochondria

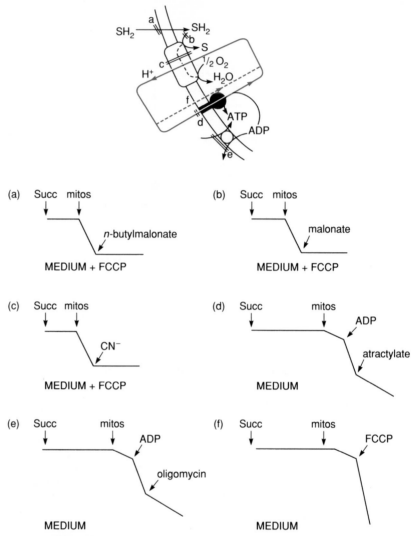

Figure 4.12   The use of the oxygen electrode to study mitochondrial energy transduction.
In the scheme six ways of interfering with energy transduction are shown. The diagrammatic oxygen electrode traces show how these perturbations might be investigated with the oxygen electrode. The incubation medium is presumed to contain osmotic support, a pH buffer and Pi.

(a) Inhibition of a transport protein (in this case for succinate)
(b) Inhibition of a substrate dehydrogenase (again for succinate)
(c) Inhibition of a respiratory chain complex (by cyanide)
(d) Inhibition of adenine nucleotide transport across the inner membrane by atractylate
(e) Inhibition of the ATP synthase by oligomycin
(f) Removal of respiratory control by the protonophore FCCP.

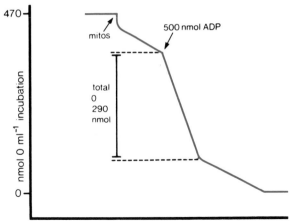

Figure 4.13   Respiratory 'states' and the determination of P/O ratios.
In this experiment mitochondria were added to an oxygen-electrode
chamber, followed by succinate as substrate. Respiration is slow as the
proton circuit is not completed by proton re-entry through the ATP synthase. That there is any respiration at all is because of a slow proton leak
across the membrane. A limited amount of ADP is added-allowing the
ATP synthase to synthesize ATP coupled to proton re-entry across the
membrane. When this is exhausted, respiration slows and finally anoxia is
attained. The circled numbers refer to the respiratory 'states'. If the
amount of ADP is known, the oxygen uptake during the accelerated state
3 respiration can be quantified allowing a P/O ratio to be calculated (moles
ATP synthesized per mol O). Since the proton leak is almost negligible in
state 3 the total oxygen uptake during state 3 is effectively used for ATP
synthesis. In this example the ADP/O ratio for the substrate is found to
be $500/290 = 1.72$. Note the bioenergetic convention of referring to 'O',
i.e. $\frac{1}{2}O_2$, which is equivalent to $2e^-$. Also, the controlled respiration prior
to addition of ADP, which is strictly termed 'state 2,' is functionally the
same as state 4, and the latter term is usually used for both states.

contain adenine nucleotides within their matrices, the amount is relatively
small (about 10 nmol (mg protein)$^{-1}$), and when the mitochondria are introduced into the incubation this pool will be very rapidly phosphorylated until
it achieves equilibrium with Δp. That there is any respiration at all is only
because the inner membrane is not completely impermeable to protons, which
can therefore slowly leak back across the membrane even in the absence of net
ATP synthesis. One factor which contributes to this proton leak is the slow
cycling of Ca^{2+} across the membrane (Section 8.3) although mostly it is due
to an endogenous 'non-ohmic' proton leak which only becomes apparent at
very high Δp (Section 4.9.1a). Only the terms 'state 3' and 'state 4' continue
to be commonly used in the 1990s.

How does the respiratory chain know how fast to respire? The fundamental
factor which actually controls the rate of respiration is the thermodynamic disequilibrium between the redox potential spans across the proton-translocating
regions of the respiratory chain and Δp. In the absence of ATP synthesis respiration is automatically regulated so that the rate of proton extrusion by the

respiratory chain precisely balances the rate of proton leak back across the membrane. If proton extrusion were momentarily to exceed the rate of re-entry, Δp would increase, the disequilibrium between the respiratory chain and Δp would in turn decrease and respiratory chain activity would decrease, restoring the steady state. Once again the electrical circuit analogy is useful here.

In the example shown in Fig. 4.13 respiration is disturbed by the addition of exogenous ADP, mimicking an extra-mitochondrial hydrolysis of ATP such as would occur in an intact cell. The added ADP exchanges with matrix ATP via the adenine nucleotide translocator, and as a result, the ΔG_p for the ATP synthesis reaction in the matrix is lowered, disturbing the ATP synthase equilibrium. The following events then occur sequentially (but note that the gaps between them would be on the millisecond timescale):

(a) The ATP synthase operates in the direction of ATP synthesis and proton re-entry to attempt to restore the ΔG_p.
(b) The proton re-entry lowers Δp.
(c) The thermodynamic disequilibrium between the respiratory chain and Δp increases.
(d) The proton current and hence respiration increases.

This accelerated 'State 3' respiration is once more self-regulating so that the rate of proton extrusion balances the (increased) rate of proton re-entry across the membrane. Net ATP synthesis, and hence state 3 respiration, may be terminated in three ways:

(a) when sufficient ADP is phosphorylated to ATP for thermodynamic equilibrium between the respiratory chain and Δp to be regained;
(b) by preventing adenine nucleotide exchange across the membrane with an inhibitor such as atractylate (Section 8.4.3);
(c) by inhibiting the ATP synthase, for example by the addition of oligomycin (Section 7.2).

Energy transduction between the respiratory chain and the protonmotive force is extremely well regulated, in that a small thermodynamic disequilibrium between the two can result in a considerable energy flux. Thus Fig. 4.14 shows that Δp drops by less than 30% when ADP is added to induce maximal state 3 respiration. The actual disequilibrium between the respiratory chain and Δp is even less, as the ΔE values across proton translocation segments of the respiratory chain also decrease in state 3.

Effective energy transduction during state 3 is also apparent at the ATP synthase. A high rate of ATP synthesis can be maintained with only a slight thermodynamic disequilibrium between Δp and ΔG_p.

Proton translocators uncouple oxidative phosphorylation by inducing an artificial proton permeability in bilayer regions of the membrane (Section 2.3.5a). They may thus be used to override the inhibition of proton re-entry which results from an inhibition of net ATP synthesis. As a consequence

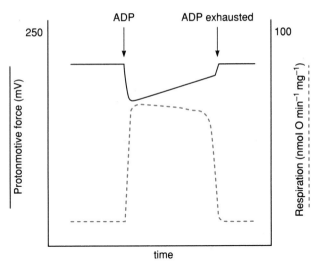

Figure 4.14   A slight fall in Δp allows a large increase in respiration. The components of Δp were measured for rat liver mitochondriac oxidizing β-hydroxybutyrate before, during and after a small addition of ADP to allow state 3 respiration. Note that a large enhancement of respiration is caused by a modest drop in Δp. Based upon data from Locke *et al.* (1982).

proton translocators such as FCCP can induce rapid respiration, regardless of the presence of oligomycin or atractylate, or the absence of ADP (Fig. 4.12).

4.7 PROTON CONDUCTANCE

In an electrical circuit, the conductance of a component is calculated from the current flowing per unit potential difference. A similar calculation for the proton circuit enables the effective proton conductance of the membrane ($C_M H^+$) to be calculated:

$$C_M H^+ = J_{H^+}/\Delta p \qquad [4.4]$$

This can be illustrated by a simple example.

Liver mitochondria oxidizing succinate in 'State 4' might typically respire at 15 nmol O min^{-1} mg^{-1} and maintain a Δp of 220 mV. If the H$^+$/O ratio for the span succinate–oxygen is 6, then:

$$J_{H+} = (\delta O/\delta t) \times (H^+/O) = 15 \times 6 = 90 \text{ nmol H}^+ \text{ min}^{-1} \text{ mg}^{-1}$$

$$C_M H^+ J_{H^+}/\Delta p = 90/220 = 0.4 \text{ nmol H}^+ \text{ min}^{-1} \text{ mg}^{-1} \text{ mV}^{-1}$$

If now the protonophore FCCP is added and Δp drops to 40 mV while respiration increases to 100 nmol O min^{-1} mg^{-1}, then the new values are as follows:

$$J_{H^+} = (\delta O/\delta t) \times (H^+/O) = 100 \times 6 = 600 \text{ nmol H}^+ \text{ min}^{-1} \text{ mg}^{-1}$$

$$C_M H^+ J_{H^+}/\Delta p = 600/40 = 15 \text{ nmol H}^+ \text{ min}^{-1} \text{ mg}^{-1} \text{ mV}^{-1}$$

The magnitude of the endogenous proton conductance of the membrane is the parameter which underlies the bioenergetic behaviour of a given preparation of mitochondria. Evidently, for an efficient transduction of energy, $C_M H^+$ should be as low as possible.

The respiratory chain would not be expected to distinguish between a Δp which is lowered by the addition of a proton translocator and one which is lowered by ATP synthesis. Any significant discrepancy would indicate either that a 'localized proton circuit' between a respiratory chain complex and an ATP synthase existed which was not accessible to the proton translocator or that a methodological error existed in the determination of one of the parameters. This is one of the areas of the chemiosmotic theory where there is continuing debate: we shall first discuss the simplest case, where no discrepancies are observed, and then analyse some controversial experiments which have led some workers in the field to suggest that the simple 'delocalized' concept is inadequate.

4.7.1 Brown adipose tissue and the analysis of the proton circuit

Review   Nicholls and Locke, 1984

Brown adipose tissue is the seat of *non-shivering thermogenesis*, the ability of hibernators, cold-adapted rodents and new-born mammals in general to increase their respiration and generate heat without the necessity of shivering. In extreme cases whole-body respiration can increase up to 10-fold as a result of the enormous respiration of this tissue, which rarely accounts for more than 5% of the body weight even in a small rodent. The tissue is innervated by noradrenergic sympathetic neurones, and release of the transmitter onto an unusual class of β-receptors activates adenylyl cyclase and hence hormone-sensitive lipase. This leads to the hydrolysis of the triglyceride stores which are present in multiple small droplets, giving the cell a 'raspberry' appearance. The brown adipocytes are packed with mitochondria, whose extensive inner membranes indicate a high capacity for respiration. However, the chemiosmotic theory now poses a problem: how can the fatty acids liberated by lipolysis be oxidized by the mitochondria when the main contributor to rate limitation in the proton circuit is the re-entry of protons into the mitochondrial matrix? The problem is compounded by the relatively low amount of ATP synthase and by the absence of any significant extra-mitochondrial ATP hydrolysis activity.

Two solutions are possible from first principles: either the brown fat mitochondrial respiratory chain is modified so that it does not pump protons, or the membrane is modified to allow re-entry of protons in the absence of ATP synthesis. The latter turns out to be the case. The mitochondrial inner membrane contains a unique 32 kDa *uncoupling protein* which binds a purine nucleotide to its cytoplasmic face and is inactive until the free fatty acid concentration in the cytoplasm starts to rise. The protein then binds a fatty acid and alters its conformation to become proton conducting (probably hydroxyl ions are the transported species). The uncoupling protein thus acts as a self-regulating endogenous uncoupling mechanism which is automatically activated in response to lipolysis, allowing uncontrolled oxidation of the fatty acids. The low-conductance state is restored when lipolysis is terminated, and the mitochondria oxidize the residual fatty acids.

Expression of the uncoupling protein occurs in response to the adaptive status of the animal: it is present in the mitochondria at high concentration at birth, but is then repressed so that the mitochondria lose the protein and the capacity for non-shivering thermogenesis. Cold-adaptation or, interestingly, over-feeding, can, under certain conditions, lead to re-expression of the protein.

Brown adipose tissue mitochondria are particularly well suited to studies of proton circuit regulation since they have an active s,n-glycerophosphate dehydrogenase whose substrate binding site is on the outer face of the inner membrane. This means that changes in $\Delta\psi$, ΔpH or matrix volume will not affect the transport of the substrate to the enzyme, or the activity of the enzyme itself.

The proton conductance $C_M H^+$ can be increased by activating the uncoupling protein, by addition of a synthetic protonophore such as FCCP or by addition of ADP. In each case an identical relationship is obtained between the rate of controlled respiration and the measured Δp (Fig. 4.15), consistent with a simple, delocalized proton circuit. Once again only a slight decrease in

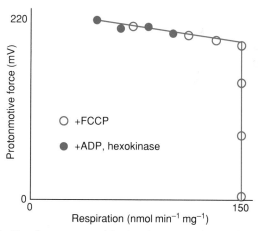

Figure 4.15   Respiratory rate of brown fat mitochondria as a function of Δp does not depend on how the latter is varied.
Mitochondria from brown adipose tissue were incubated with s,n-glycerophosphate as substrate. Δp and hence respiration were varied by increasing additions of the protonophore FCCP (○) or addition of ATP plus varying hexokinase to allow increasing state 3 respiration (●). Note that the points fall on the same line, indicating a unique relationship between respiration and Δp and also that only a slight decrease in Δp is sufficient for full uncontrolled respiration. Brown adipose tissue mitochondria have only a limited ATP synthase activity and so ADP causes less maximal respiratory stimulation than does FCCP.

Δp is required to increase respiration to the extent that the dehydrogenase becomes the principal determinant of the (uncontrolled) respiration. At this stage Δp is still greater than 150 mV.

4.8 THE MITOCHONDRIAL RESPIRATION RATE CAN BE CONTROLLED BY SEVERAL FACTORS

Review   Brand and Murphy, 1987

In the previous section we have explained the connections between the mitochondrial respiration rate, Δp and $C_M H^+$. Here we consider whether this is the only factor that is important in establishing the mitochondrial respiration rate. In recent years it has become apparent that assigning the control of a metabolic pathway to a single reaction step is an oversimplification. Several steps may contribute to the control, and a quantitative *control analysis* has been developed to provide a simple description of how control is distributed.

We can illustrate control analysis with a simple example. Consider a mitochondrion respiring in state 3. If we deliberately alter the activity of a single step in the overall sequence, for example the adenine nucleotide translocator, by a small fraction, say 1%, what effect does this have on the overall respiration rate? Two extreme results are possible in this type of experiment. First, the change in flux through the entire pathway may be the same percentage as the change in activity of the single step, i.e. 1%. In this case the *flux control coefficient* of the adenine nucleotide translocator would be said to be 1 and, under the particular set of conditions analysed, the other enzymes and transporters in the sequence would not contribute to the control, with the corollary that their flux control coefficients would be 0. The second extreme would be when a 1% change of the translocator activity had no effect on the overall flux. In this case the step would have a flux control coefficient of 0.

In practice flux control coefficients of 1 are rare; the idea of a single rate-determining step, to which a flux control coefficient of 1 corresponds, although often encountered in chemical reactions, probably rarely applies to metabolic sequences. Instead there is an interplay between many steps, each of which may have significant flux control coefficients, with values in non-branched pathways between 0 and 1. In any pathway the sum of all the individual flux control coefficients is always 1. Mitochondrial oxidative phosphorylation is especially suited to this type of analysis because specific inhibitors are available for a range of steps. Because there are no specific inhibitors for the proton leak, its flux control coefficient is calculated indirectly from its kinetic response to Δp under specified conditions; this response is known as the elasticity of the leak to Δp.

With the above background we shall consider the factors that control the respiration of a suspension of mitochondria during a transition from state 4 to state 3. The mitochondria are supplied with succinate as respiratory substrate together with ADP and Pi. The initial state 4 is attained when the net conversion of ADP and Pi to ATP ceases. As intuitively expected, the overall flux control coefficient of the set of reactions adenine nucleotide translocation, ATP synthesis and consumption of ATP ('ATP turnover') are 0 in state 4. In the previous sections we have made the simplification that the proton leak across the mitochondrial membrane completely controls the respiration in state 4 (i.e. has a flux control coefficient of 1). However, more careful analysis shows that although the proton leak is indeed dominant (flux control coefficient 0.9) there is also significant control (coefficient of 0.1) in the set of reactions (called here 'succinate utilization') catalysing transport of succinate into the mitochondrion and its oxidation by the electron-transport chain.

If glucose and incremental amounts of hexokinase are now added, respiration will steadily increase until the rate of ATP synthesis reaches a maximum and the mitochondria are in state 3. The first additions of hexokinase each cause a marked increase in the respiration rate and thus the flux control coefficient of the 'ATP turnover' reactions is high, corresponding to the classic respiratory control (Fig. 4.16). As further hexokinase is added, other components of the 'ATP turnover' reactions, particularly the adenine nucleotide translocator, assume an increasing share of the control. Concomitantly the control by hexokinase becomes a progressively smaller component of the control exerted by the 'ATP turnover' reactions. At the limit of state 3 respiration further additions of hexokinase are without effect on the respiration rate and thus its flux control coefficient falls to 0, the classic 'uncontrolled respiration' (state 3).

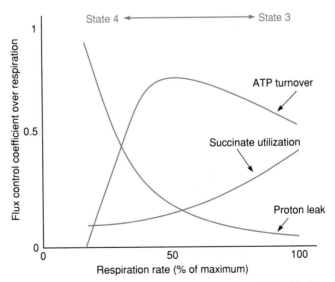

Figure 4.16 Control strength of steps in mitochondrial oxidative phosphorylation during a progressive transition from state 4 to state 3.
As described in the text, as the respiration rate increases from state 4 to state 3 with increasing addition of hexokinase to increase the rate of the ATP turnover reactions, the flux control coefficients alter as shown (adapted from Hafner *et al.* (1990), courtesy of Dr M.D. Brand).

In state 3, control is shared almost equally (Fig. 4.16) between 'ATP turnover' reactions and those of 'succinate utilization'. More detailed analysis (Groen *et al.*, 1983) shows that it is distributed between the adenine nucleotide translocator, the dicarboxylate translocator, the cytochrome bc_1 complex and cytochrome oxidase. As the respiration rate alters between the extremes of state 3 and state 4 the quantitative contribution of each of these components varies; for example, control due to the adenine nucleotide translocator rises to its greatest flux control strength at 75% of the maximum respiration rate. The important outcome of this analysis is that in neither state 3 nor state 4 is a single step responsible for the control of the mitochondrial respiration rate. Traditional attempts to correlate respiratory control with the [ATP]/ [ADP] [Pi] ratio or a single irreversible step in the electron-transport chain (e.g. a step in the cytochrome oxidase reaction) appear no longer to be tenable.

In an intact cell the factors controlling mitochondrial respiration rate will be more varied and complex than those considered above. The major respiratory substrate will not be succinate but rather NADH generated in the matrix. There will also be important differences between mitochondria from different cell types. Mitochondria in a liver cell respire at a rate intermediate between state 3 and state 4. Control analysis shows that this rate is controlled by processes (such as glycolysis, fatty acid oxidation and the citric acid cycle) that supply mitochondrial NADH (flux control coefficient 0.15 to 0.3) by the proton leak (flux control coefficient 0.5) and by the 'ATP turnover' reactions (flux control coefficient 0.5). Oxidation of NADH is less important with a flux control coefficient between 0 and 0.15. Fluctuations in rate can be caused by hormones or increases in cytoplasmic and matrix Ca^{2+} via three separate effects; alteration of either ATP turnover, NADH supply or proton leak. Each of these effects may be important. Muscle mitochondria can experience periods of resting activity when they may be close to the state 4 respiration rate but upon initiation of contraction the ATP demand and raised Ca^{2+} may be such as to cause transition to state 3. If anaerobiosis approaches, the rate of respiration could conceivably pass transiently through a stage where cytochrome oxidase has a higher flux control coefficient owing to restriction on the supply of oxygen.

The control of the electron-transport rate in bacteria and thylakoids can, in principle, be understood in similar terms but less experimental effort has been expended on these systems. Usually, intact bacterial cells respire at their maximum rate; there is no stimulation on adding a protonophore, i.e. the proton leak has a very low flux control coefficient. However, the distribution of control between the electron-transfer chains, ATP synthase, protonsymports for nutrients and other cellular processes is unknown. Electron transport in thylakoids is, in common with mitochondria, restricted when net ATP synthesis is not occurring, i.e. the proton leak will have significant control under these conditions.

4.9 CONTROVERSIES

Although discussion of the structure and mechanism of respiratory chain and ATP synthase proton pumps is reserved for later chapters, it is necessary at this stage to review some of the controversies which have arisen from a detailed investigation of the quantitative aspects of the proton circuit, particularly since much of the research effort since the publication of the first edition of this book has been directed towards such an analysis. The controversies can be divided into two classes: those which concern the mechanism by which the proton pumps operate and those which relate to the proton circuit *per se* and the coupling of respiration to the ATP synthase. In the view of the authors the former do not compromise the chemiosmotic theory, whereas the latter, if substantiated, would demand a dramatic reassessment or even abandonment of the theory. It is our opinion that this stage has yet to be attained.

4.9.1 Controversies which relate to proton pump mechanism

Since the structural details of the proton pumps will be covered in subsequent chapters we shall restrict ourselves to general aspects which arise during investigation of the proton circuit. Detailed discussion of stoichiometries of individual complexes and 'loop' versus 'conformational pump' debates will be reserved for Chapter 5.

(a) *The non-linear relation between Δp and state 4 respiration: non-ohmic proton leak or molecular slip?*

Selected references   Zolkiewska *et al.* 1989, Duszynski and Wojtczak, 1985, Pietrobon *et al.* 1986, Zoratti *et al.* 1986, Wojtczak *et al.* 1990, Nobes *et al.* 1990

However carefully they are prepared, all mitochondria show a significant state 4 respiration. This endogenous proton leak shows an interesting dependency on the magnitude of Δp which is revealed by varying the substrate supply and monitoring Δp and respiration (Fig. 4.17). At high Δp, particularly in the presence of flavoprotein-linked substrates, a large increase in the apparent proton conductance of the membrane is seen. The function of this 'non-ohmic leak' may be to act as a safety valve limiting the potential across the membrane and preventing the risk of dielectric breakdown: after all, the membrane is maintaining a potential of <200 mV across a lipid bilayer of some 7 nm, corresponding to a field strength of 300 000 V cm^{-1}! The proton leak may also contribute significantly to the resting respiratory rate of intact cells.

An alternative explanation for this non-linear relationship between state 4 respiration and Δp at high potentials is that there may be intrinsic uncoupling at the level of the respiratory chain proton pumps themselves such that some catalytic cycles of the complexes might be able to occur in the absence of stoichiometric proton translocation across the membrane. There is nothing inherently unlikely about such a 'slip' mechanism, which would not require any modification of the delocalized concept of the proton circuit. Nevertheless it is a contentious issue, with research groups roughly evenly divided between

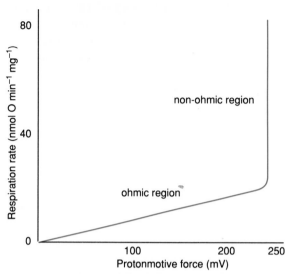

Figure 4.17 The non-ohmic proton leak in mitochondria.
Brown adipose tissue mitochondria were incubated in state 4 (in the presence of excess GDP to inhibit the uncoupling protein and increasing concentrations of *sn*-glycerol-3-phosphate as substrate. Respiration increased
with substrate concentration although Δp plateaued. This indicates that
the proton conductance of the membrane C_MH^+ increases when Δp rises
to a critical value, the so-called 'non-ohmic leak'.

those favouring 'slips', i.e. involving the mechanism of the complexes themselves, and non-ohmic 'leaks'. One problem is that it is not easy to determine
C_MH^+ directly, e.g. by monitoring the kinetics of the collapse of the proton
gradient when proton pumping is inhibited, since Δp will rapidly drop below
the non-ohmic threshold, although it is possible to detect a rapid initial fall in
$\Delta \psi$ when respiration is suddenly inhibited, consistent with a high non-ohmic
proton conductance. Non-ohmic behaviour may be an inherent property of the
phospholipid bilayer (Krishnamoorthy and Hinkle, 1984), contributing to the
desirable homeostasis of the protonmotive force, not only in mitochondria but
also in bacterial and thylakoid membranes.

(a) *Variable stoichiometry of proton pumps*

As discussed in Chapter 3, equilibrium thermodynamics can be used to determine the stoichiometry of an energy transduction. If conditions are arranged
such that the proton-translocating ATPase is at equilibrium, determination of
the magnitude of ΔG for the ATPase reaction and Δp allows the H^+/O stoichiometry to be calculated. It would naturally be expected that this ratio would
remain constant at a variety of values of Δp. However, there are several
reports of changes in this apparent stoichiometry (e.g. Murphy and Brand,
1987, 1988; Azzone *et al.*, 1985) and the question has to be answered whether
this represents a true mechanistic shift in stoichiometry or an experimental

artifact. There is a severe conceptual problem with a variable stoichiometry; just like a car changing gear a decrease in, for example, the $H^+/2e^-$ stoichiometry of a respiratory chain complex from 3 to 2 would allow a 50% higher Δp to be attained at equilibrium or, if this were not possible, would ensure that the complex became essentially irreversible. Neither of these is seen in practice and a recent paper (Wojtczak *et al.*, 1990) suggests reasons for previous contradictory results.

4.9.2 Controversies concerning the proton circuit

Review Ferguson 1985

A number of experiments have been proposed to show anomalous relationships in the kinetics and thermodynamics of the proton circuit, most of which, if substantiated, would require a fundamental re-evaluation of the simple 'delocalized' proton circuitry which forms the basis of the chemiosmotic theory as described in this book. Each of the following observations is controversial in the literal sense that other research groups have disputed the findings in favour of a delocalized interpretation. In the view of both the present authors none of the anomalies are proven, although rigorous final analysis is not available in all cases.

(a) *Is there a class of membrane active agents which can stimulate respiration without lowering Δp?*

Selected references Rottenberg 1986, Schonfeld *et al.* 1989

The action of protonophores, 'uncouplers', on the proton circuit should be relatively straightforward: an increase in proton conductance causes a lowering of Δp and a relaxation of respiratory control. Nevertheless, there have been reports that some fatty acids, gramicidin derivatives and local anaesthetics increase state 4 respiration without a detectable lowering of Δp. These agents have been termed 'decouplers' (Rottenberg, 1986). These results are not readily explained by delocalized chemiosmosis since they suggest that the respiratory chain complexes can in some way be locally 'short-circuited' in a way which does not involve the proton circuit or affect the measurable proton conductance of the membrane (Fig. 4.18). However, in the 'delocalized' paradigm only a slight decrease in Δp appears necessary to produce a large stimulation of respiration (Figs 4.14, 4.15). Before 'decouplers' become generally accepted it would be necessary to eliminate possible artifacts such as a 'decoupler'-induced swelling of the mitochondrial matrix which might distort the calculations of Δp since these require an accurate knowledge of the matrix volume (Fig. 4.3). It is notable also that other research groups find that the proposed 'decoupler' long-chain fatty acids behave as typical protonophores, acting to increase $C_M H^+$.

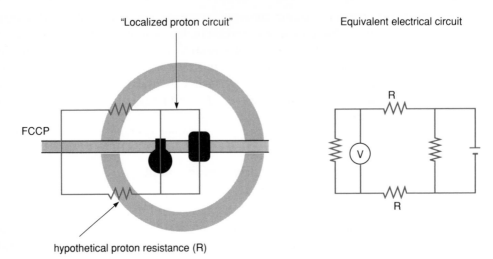

Figure 4.18 Localized and delocalized proton circuits.
'Localized' theories propose that there are local circuits between individual respiratory chain assemblies and ATP synthases. There would be a signifi-cant 'resistance' to be crossed for protons leaving this microdomain, which would be detected as a discrepancy in the respiratory stimulation corre-sponding to a given depression in Δp (detected outside the domain) when FCCP was added or when ADP was added: Δp would have to drop further with FCCP than with ADP (see text).

(b) *Does the relationship between the rate of ATP synthesis and Δp depend on whether the latter is varied by inhibition of electron transport or by partial uncoupling?*

Selected references   Sorgato *et al.* 1985, Davis and Davis van Thienen 1989

The essence of the chemiosmotic theory is that the ATP synthase should not sense changes in respiratory chain activity directly but rather through alter-ations in the steady-state Δp. Consequently if Δp were lowered to the same extent, either by adding a small amount of protonophore or by restricting the rate of electron transport with an inhibitor (or in the case of a photosynthetic system by lowering the light intensity), then the resultant changes in the rate of ATP synthesis would be the same. A number of observations conflicted with this expectation, suggesting that the ATP synthase was able to respond directly to changes in the activity of the electron-transport chain. However, there are experimental difficulties in the measurement of Δp under these two conditions and at the time of writing (autumn 1991) it appears that any discrepancy may be artifactual.

(c) *Does the equilibrium relationship between ΔG_p and Δp remain the same when protonophore is titrated in?*

Selected reference   Woelders *et al.* 1985

A number of early reports suggested that ATP could still be synthesized by mitochondria when sufficient protonophore was present to decrease Δp to low values. The discrepancy was too large to be realistically accounted for by a hypothetical increase in the H^+/ATP stoichiometry of the ATP synthase (Section 2.5), and instead a *localized proton circuit* was proposed, in which individual respiratory chains were in intimate relationship with individual ATP synthases and linked by isolated proton circuits (Fig. 4.18). In contrast, a significant resistance to proton flow was proposed through the intermembrane space and/or matrix between the same respiratory chains and the sites at which the protonophores (and presumably the indicators of $\Delta\psi$ and ΔpH) were located. Later one of the original proponents of this localized circuit (Woelders *et al.*, 1985) pointed out three possible artifacts in the original experiments: that the maximum ΔG_p was underestimated due to difficulties in quantifying a very low ADP concentration, that the minimum ΔG_p was exaggerated due to artifactual ATP formation via adenylate kinase, and that Δp in the presence of protonophore could collapse artifactually in the anaerobic pellet required by the procedure. More recent reports show no discrepancies in either the equilibrium relationship or the kinetic relationship discussed in the previous section, see also Fig. 4.15.

4.9.3   How do alkaliphilic bacteria make ATP?

Review   Krulwich *et al.* 1988

Alkaliphilic bacteria grow using aerobic respiration in environments where the pH is as high as 12. Since the pH of their cytoplasm is between 8 and 9, this means that the ΔpH across their cytoplasmic membrane can be as much as 3 pH units (equivalent to 180 mV) acidic inside, i.e. in the opposite sense to that required to contribute to the Δp. Unless the bacteria were to maintain an exceptionally high $\Delta\psi$ of the order of 400 mV to compensate for the reversed pH gradient, it is difficult to see how they could synthesize ATP by a conventional mechanism. In the most thoroughly investigated systems, *Bacillus firmus* and *Bacillus alcalophilus*, $\Delta\psi$ is of the order of 180 mV, positive outside. Thus the total Δp appears to be in the range of 5–80 mV; the exact value depends on the organism and is smaller the higher the external pH.

How then is ATP synthesized in these bacteria? Several proposals have been made. The first is that the proton circuit would occur across internal vesicle membranes and thus be independent of the Δp across the cytoplasmic membrane. Attractive though this idea might be, no evidence for such a mechanism has ever been obtained. A second proposal is that ATP synthesis could be driven by a *sodium* circuit rather than a proton circuit. In some bacteria such as the marine *Vibrio* marginal alkaliphiles this mechanism seems to apply. However, in the *Bacillus* species discussed above there is no primary Na^+ pumping, although primary proton pumping does generate a secondary Na^+ electrochemical gradient via a Na^+/H^+ antiporter. This is used for solute accumulation via a variety of Na^+:solute symporters but not for ATP synthesis. The proton re-entry coupled to this secondary Na^+ cycle is also important for pH regulation. The electron-transport chain translocates protons and the ATP synthase is a Na^+-independent H^+-translocating F_0F_1 ATP synthase.

What possibilities remain? If the proton circuit across the cytoplasmic membrane really does drive ATP synthesis, the relationship between Δp and ΔG_{ATP} for the ATP synthase (Chapter 3) would require an extremely high H^+/ATP stoichiometry. The low Δp would similarly imply a high H^+/O ratio for effective conservation of the redox potential energy but such high values have not been observed. However, it is also pertinent to recall (Section 4.2.1) that the methods for determining the components of Δp are indirect and subject to error. Thus at the time of writing Hoffmann and Dimroth (1991) have argued that for *B. alcalophilus* at external pH = 10, the $\Delta\psi$ is as much as 210 mV and the total Δp around 105 mV, consistent with a chemiosmotic mechanism provided that H^+/ATP is 4 or 5. This does not necessarily explain the energy-transduction process (independent verification of this stoichiometry would be needed, see Section 4.2.5) and does not account for the circumstance when the external pH is as high as 12. A more revolutionary idea would be that in this particular case the chemiosmotic mechanism does not apply to these organisms and that some direct 'localized' proton transfer from respiratory chain to ATP synthase could occur (Fig. 4.18).

4.10 OVERALL PARAMETERS OF ENERGY TRANSDUCTION (P/O RATIOS)

4.10.1 Respiratory control ratio

This is an empirical parameter for assessing the integrity of a mitochondrial preparation. It is defined as the respiratory rate attained during maximal ATP synthesis (i.e. in the presence of ADP), or in the presence of a proton translocator, divided by the rate in the absence of ATP synthesis or proton translocator. It is therefore a hybrid parameter depending on a number of primary parameters, notably the endogenous proton leak (Section 4.9.1a). Typical values for the ratio vary from 3 to 15 in different preparations. Note that most bacterial cells do not show significant respiratory control owing, presumably, to the continuous activity of the ATP synthase and other protonmotive force-driven reactions; respiratory control can be observed in some inside-out vesicle preparations. The rate of electron transport in isolated thylakoids does accelerate when ATP synthesis is occurring.

4.10.2 P/O (ATP/O, ADP/O) and P/2e⁻ (ATP/2e⁻, ADP/2e⁻) ratios

Review Ferguson 1986

While the stoichiometries of proton translocation by the respiratory chain and ATP synthase appear to be fixed, even if the actual values are still a matter for contention, the overall stoichiometry of mitochondrial ATP synthesis in relation to respiration can vary from a theoretical maximum of about 2.5 ATP per 2 e$^-$ passing from NADH to oxygen down to zero, depending on the activity of the parallel proton leak pathway bypassing the ATP synthase. The variety of terms in the heading are used to describe this empirical ratio; they are roughly synonymous.

The P/O ratio is the number of moles of ADP phosphorylated to ATP per 2 e$^-$ flowing through a defined segment of an electron transfer to oxygen. If the terminal acceptor is not oxygen then the term P/2 e$^-$ ratio is used. P/O ratios can be determined from the extent of the burst of accelerated state 3 respiration obtained when a small measured aliquot of ADP is added to mitochondria respiring in state 4, (Fig. 4.13). Almost all the added ADP is phosphorylated to ATP, the ATP/ADP ratio being typically at least 100:1 when state 4 is regained, and the ratio 'moles of ADP added/moles of O consumed' can be calculated. It is the convention to assume that the proton leak ceases during state 3 respiration, which is largely true, due to its non-ohmic nature (Fig. 4.17), and thus the total O consumed should be used in the calculation.

Values for ADP/2e$^-$ ratios are, as with all stoichiometries, a source of debate. The 'classic' value of 1 ATP per 2 e$^-$ per proton-translocating complex is no longer tenable. For example, complex III (UQH$_2$-cyt c) has a H$^+$/2 e$^-$ ratio of 4, but because electrons enter the complex from the matrix side and leave from the cytoplasmic face, the charge/2 e$^-$ (q$^+$/2e$^-$) ratio is only 2 (see Chapter 5). Conversely, complex IV (cyt c–O$_2$) has a H$^+$/2e$^-$ ratio of 2 but a q$^+$/2e$^-$ ratio of 4 (Chapter 5). To maintain overall electroneutrality, positive charges, in the form of protons, must flow back through the ATP synthase to balance the charge displacement, and to make ATP it follows that the P/2e$^-$ ratio for complex IV should be twice that for complex III. This is more than a semantic argument for many bacteria, since electron transfer frequently terminates at the level of cyt c and so a dissection of P/2e$^-$ ratios for individual parts of an electron-transport system is important.

4.10.3 Non-integral P/O ratios

Taking the consensus view for mitochondria that the H$^+$/2e$^-$ ratio is 6 for the span succinate–O$_2$, that the H$^+$/ATP ratio at the ATP synthase is 3, and that one additional proton is consumed in the transport of Pi and translocation of adenine nucleotides, it follows that the theoretical maximum P/O ratio for succinate oxidation would be 1.5 (i.e. 6 ÷ 4), close to what is observed, rather than the 'classic' value of 2 (Hinkle et $al.$, 1991). There is still some uncertainty about the H$^+$/2e$^-$ stoichiometry of complex I (NADH–UQ) but it could be as high as 4 and thus H$^+$/O for the span NADH–O$_2$ would be 10. If so then the maximum P/O ratio for this span would be 2.5 (10 ÷ 4). P/O values of 2.5 and 1.5, rather than the traditional 3 and 2, for mitochondrial oxidation of NADH and succinate (or of other substrates from which electrons enter the chain at the level of ubiquinone) mean that the standard textbook stoichiometries of ATP synthesis associated with the total oxidation of carbohydrates and fats require to be revised downwards. Similar considerations apply to ATP synthesis in bacterial and thylakoid systems; the maximum P/2e$^-$ ratio will be determined by the relative values of H$^+$/2e$^-$ and H$^+$/ATP. Note that the translocation step for adenine nucleotide and phosphate is not involved in the synthesis of ATP by bacterial and thylakoid membranes, nor indeed in SMPs, and thus these systems should therefore have higher P/2e$^-$ ratios for given H$^+$/2e$^-$ and H$^+$/ATP (for the ATP synthase) stoichiometries.

Non-integral ratios present problems for hypothetical 'localized' proton circuits where individual respiratory chains are envisaged as being tightly coupled to individual ATP synthase complexes (Fig. 4.18). The important point to recognize here is that the chemiosmotic mechanism of energy coupling, being an indirect and delocalized mechanism, does not require whole- number stoichiometries, unlike tightly coupled chemical reactions.

4.11 REVERSED ELECTRON TRANSFER AND THE PROTON CIRCUIT DRIVEN BY ATP HYDROLYSIS

The ATP synthase is reversible and is only constrained to run in the direction of net ATP synthesis by the continual regeneration of Δp and the use of ATP by the cell. If the respiratory chain is inhibited and ATP is supplied to the mitochondrion, the ATP synthase functions as an ATPase, generating a Δp comparable to that produced by the respiratory chain. The proton circuit generated by ATP hydrolysis must be completed by a means of proton re-entry into the matrix. Proton translocators therefore accelerate the rate of ATP hydrolysis, just as they accelerate the rate of respiration; this is the 'uncoupler-stimulated ATPase activity'.

The classic means of discriminating whether a mitochondrial energy-dependent process is driven directly by Δp or indirectly via ATP is to investigate the sensitivity of the process to the ATP synthase inhibitor oligomycin. A Δp-driven event would be insensitive to oligomycin when the potential was generated by respiration, but sensitive when Δp was produced by ATP hydrolysis. The converse would be true of an ATP-dependent event.

The near equilibrium in state 4 between Δp and the redox spans of complexes I and III suggests that conditions could be devised in which these segments of the respiratory chain could be induced to run backwards, driven by the inward flux of protons. It should be noted that this does not apply to complex IV, which is essentially irreversible. Reversed electron transfer may be induced in two ways, either through generating a Δp by ATP hydrolysis, or by using the flow of electrons from succinate or cyt c to O$_2$ to reverse electron transfer through complexes I or I and III respectively (Fig. 4.19). Such a flow of electrons, e.g. from succinate, involves the majority of the electron flux passing to O$_2$ and thereby generating Δp whilst a minority is driven energetically uphill to reduce NAD$^+$ at the expense of Δp.

Under physiological conditions the mitochondrial ATP synthase will not normally be called upon to act as a proton-translocating ATPase, except possibly during periods of anoxia when glycolytic ATP could be utilized to

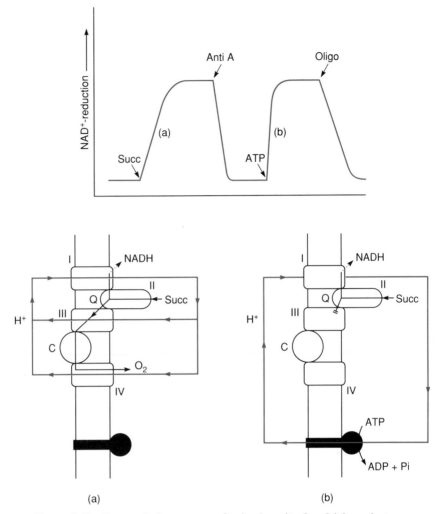

Figure 4.19 Reversed electron transfer in the mitochondrial respiratory chain.

Schematic response of submitochondrial particles incubated in the presence of NAD^+. In (a), a Δp is generated by succinate oxidation. Δp then drives complex I in reverse, causing NAD^+ reduction, i.e. succinate acts as both donor of electrons for reversed electron transfer and as substrate for complexes III and IV. In (b) complex III is inhibited by antimycin A and the Δp is generated by ATP hydrolysis. Succinate merely donates electrons for reversed electron transfer through complex I.

maintain the mitochondrial Δp. However, some bacteria, such as *Streptococcus faecalis* when grown on glucose, lack a functional respiratory chain and rely entirely upon hydrolysis of glycolytic ATP to generate a Δp across their membrane and enable them to transport metabolites. Reversed electron transport driven by Δp generated through respiration is an essential process in some bacterial species (see Chapter 5).

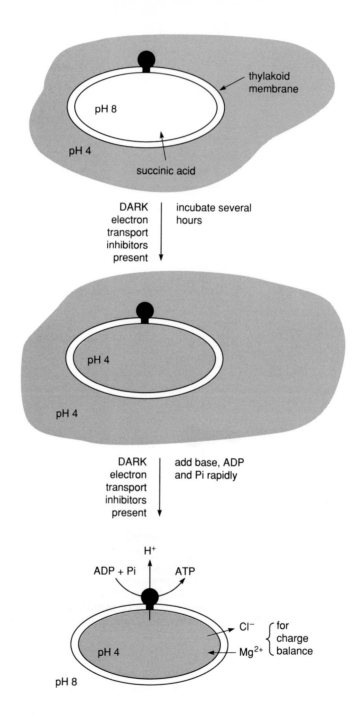

Figure 4.20   The 'acid bath' experiment: a ΔpH can generate ATP. Thylakoid membranes were incubated in the dark at pH 4 in a medium containing succinate, which slowly permeated into the thylakoid space, liberating protons and lowering the internal pH to about 4. The external pH was then suddenly raised to 8, creating a ΔpH of 4 units across the membrane. ADP and Pi were simultaneously added and proton efflux through the ATP synthase led to the synthesis of about 100 moles of ATP per mole of synthase. Protonophores such as FCCP inhibited the ATP production.

4.12   ATP SYNTHESIS DRIVEN BY AN ARTIFICIAL PROTONMOTIVE FORCE

The chemiosmotic hypothesis predicts that an artificially generated Δp should be able to cause the net synthesis of ATP in any energy-transducing membrane with a functional ATP synthase. The first demonstration that this was so came from thylakoids (Jagendorf and Uribe, 1966). These authors found that thylakoids equilibrated in the dark at acid pH could be induced to synthesize ATP when the external pH was suddenly increased from 4 to 8, creating a transitory pH gradient of 4 units across the membrane, the *acid bath experiment* (Fig. 4.20). This experiment is possible because thylakoids normally operate with ΔpH as the main component of Δp due to the ease with which Cl^- redistributes across the thylakoid membrane to collapse $\Delta\psi$ (Section 6.4.4), while the photosynthetic electron-transfer chain and ATP synthase are happy to operate with a highly acidic thylakoid space without loss of activity or denaturation.

For an analogous acid-bath experiment with mitochondria or bacteria, an ionophore such as valinomycin is needed to allow movement of compensating charge. SMPs, which are inverted relative to intact mitochondria, are treated with valinomycin to render them permeable to K^+, incubated at low pH in the absence of K^+ to acidify the matrix, and then transferred to a medium of higher pH containing K^+. K^+ entry creates a diffusion potential, positive inside, and this, together with the artificial ΔpH which has just been created, generates a short-lived Δp. Protons exit through the ATP synthase, generating a small amount of ATP. K^+ enters on valinomycin to maintain charge balance. Eventually the K^+ and H^+ gradients run down to the extent that ATP synthesis ceases. An analogous approach has been used to demonstrate Δp-driven secondary active transport (Section 8.5.1).

4.13   KINETIC COMPETENCE OF THE PROTONMOTIVE FORCE TO SERVE AS THE ENERGY-TRANSDUCING INTERMEDIATE

4.13.1   Proton utilization

The ability to demonstrate ATP synthesis linked to the artificial imposition of a Δp permits a further crucial test of the chemiosmotic theory. If this gradient is the intermediate between electron transport and ATP synthesis then the sudden imposition of an artificial Δp of comparable magnitude to that normally produced by the respiratory chain should lead to ATP synthesis with minimal delay and at an initial rate comparable to that seen in the natural process. Clearly, if this 'artificial' ATP synthesis were delayed and slow relative to that driven by electron transport there would be strong grounds for supposing that the proton translocation lay on a side-path.

Tests of kinetic competence have been made for both the thylakoid and SMP systems as described above, except that the protonmotive force was imposed by rapid mixing. The subsequent reaction period can be altered by varying the length of tubing between the mixing and quenching points (where the reaction is terminated by concentrated acid). In this way ATP synthesis on the millisecond timescale can be followed. In both preparations ATP synthesis was initiated with no significant lag and at initial rates comparable to those seen for the normal energy transduction. Indeed, in the case of the SMPs the onset of ATP synthesis was more rapid than following initiation of respiration.

4.13.2 Proton movements driven by electron transport

While the experiments described in Section 4.13.1 are clearly consistent with the kinetic competence of Δp as the intermediate, an important complementary test would be to show that the generation of Δp by electron transport preceded ATP synthesis. This requires a method with a high time resolution for detection of Δp. The carotenoid band shift, an indicator of membrane potential in thylakoid membranes and bacterial chromatophores (Section 4.2.2.a) has an almost instant response to an imposed membrane potential, and responded within microseconds to the initiation of light-driven electron transport initiated by a laser flash. Furthermore, the subsequent decay of the membrane potential was accelerated by the presence of ADP and Pi. As the increased decay is due to the passage of protons through the ATP synthase to make ATP, it follows that ATP synthesis occurs after the formation of $\Delta\psi$.

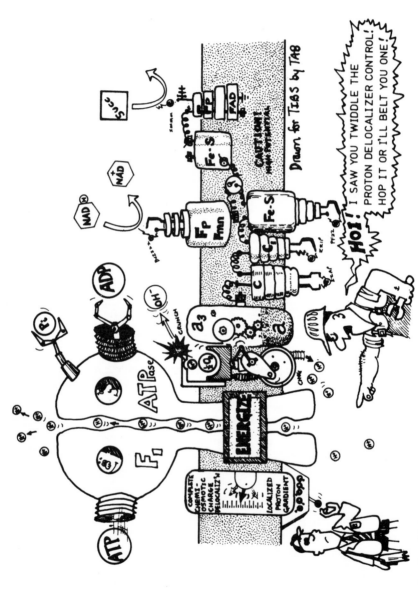

The mitochondrial respiratory chain and the ATP synthase: the localized proton circuit/delocalized chemiosmotic debate

5 RESPIRATORY CHAINS

5.1 INTRODUCTION

This chapter will look at the approaches which have been taken to investigate the structure of the respiratory chains of mitochondria and bacteria. The respiratory chain of mammalian mitochondria is an assembly of more than twenty discrete carriers of electrons which are mainly grouped into four polypeptide complexes (Fig. 5.1). Three of these complexes (I, III and IV) act as oxidation–reduction driven proton pumps. However, although the sequences of the constituent polypeptides are rapidly being elucidated, structural and functional understanding is far from complete. Only in the case of the bacterial reaction centre (Section 6.2.2) is the complete structure of an electron-transport chain known at the atomic level: in the case of the respiratory chain complexes, molecular mechanisms have to be extrapolated from the incomplete structural information available, and it is fair to say that in no case do we know the complete molecular mechanism by which protons are pumped. Despite this, and particularly in the case of the cytochrome bc_1 complex, considerable detail is available. We shall illustrate methods for studying electron transport by reference to mitochondria although comparable approaches are applied to bacteria and photosynthetic systems.

5.2 COMPONENTS OF THE MITOCHONDRIAL RESPIRATORY CHAIN

The respiratory chain transfers electrons through a redox potential span of 1.1 V, from the $NAD^+/NADH$ couple to the $O_2/2H_2O$ couple. Much of the

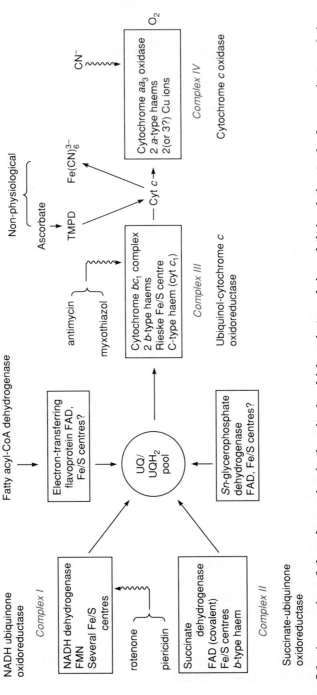

Figure 5.1 An overview of the redox carriers in the mitochondrial respiratory chain and their relation to the four respiratory chain complexes. 'Wavy arrow' = site of action of an inhibitor.

respiratory chain is reversible (Section 3.6.4), and to catalyse both the forward and reverse reactions it is necessary for the redox components to operate under conditions where both the oxidized and reduced forms exist at appreciable concentrations. In other words, the operating redox potential of a couple, E_h, (Section 3.3.3), should not be far removed from the mid-point potential of the couple, E_m. As will be shown later (Section 5.4.1), this constraint is generally obeyed, and this in turn gives some rationale to the apparently random selection of redox carriers within the respiratory chain.

The initial transfer of electrons from the soluble dehydrogenases of the citric acid cycle requires a cofactor which has a mid-point potential in the region of -300 mV and is sufficiently mobile to shuttle between the matrix dehydrogenases and the membrane-bound respiratory chain. This function is filled by the NADH/NAD$^+$ couple, which has an $E_{m,7}$ of -320 mV.

While the majority of electrons are transferred to the respiratory chain in this way, a group of enzymes catalyse dehydrogenations where the mid-point potential of the substrate couple is close to 0 mV, and they are thus not able to reduce NAD$^+$. These, succinate dehydrogenase, s,n-glycerophosphate dehydrogenase, and the 'electron-transferring flavoprotein' (transferring electrons from one of the oxidation steps in the catabolism of fatty acids by β-oxidation), feed electrons directly into the respiratory chain at a potential close to 0 mV without the intermediacy of the NAD$^+$/NADH couple (Fig. 5.1). This direct transfer requires that these enzymes be membrane-bound.

The redox carriers within the respiratory chain consist of: *flavoproteins*, which contain tightly bound FAD or FMN as prosthetic groups and undergo a $(2H^+ + 2e^-)$ reduction; *cytochromes*, with porphyrin prosthetic groups undergoing a one-electron reduction; *iron–sulphur (non-haem iron) proteins* which possess prosthetic groups also reduced in a one-electron step; *ubiquinone*, which is a free, lipid-soluble cofactor reduced by $(2H^+ + 2e^-)$; and finally *protein-bound Cu*, reducible from Cu^{2+} to Cu^+.

Cytochromes are classified according to the structure of their porphyrin prosthetic group. Mitochondria contain a-, b- and c-type cytochromes with the haem being covalently attached to the latter. Cytochromes d and o occur in some bacterial chains as terminal oxidases, o cytochromes being oxidizable b-type cytochromes, while d cytochromes possess a partially saturated chlorin ring in place of porphyrin.

5.2.1 Fractionation and reconstitution of mitochondrial respiratory chain complexes

Although the mitochondrial electron-transport chain contains approximately twenty discrete electron carriers, they do not all function independently in the membrane. The only components which fit this category are ubiquinone (also called coenzyme Q, UQ or simply Q) which is found in mammalian mitochondria as UQ_{10}, i.e. with a side chain of ten 5-carbon isoprene units (see Fig. 5.6), its reduced form ubiquinol and the water soluble cytochrome *c* that is on the P-side of the membrane.

Certain detergents (bile salts such as cholate and deoxycholate were originally used) when employed at low temperatures and low concentrations disrupt lipid–protein interactions in membranes, leaving protein–protein associations intact. Using these detergents, the mitochondrial respiratory chain can be fractionated into four complexes, termed complex I (or NADH–UQ oxidoreductase), complex II (succinate dehydrogenase), complex III (UQH$_2$–cytochrome

(a) Split-beam spectrophotometer

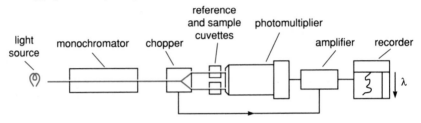

(b) Dual-wavelength spectrophotometer

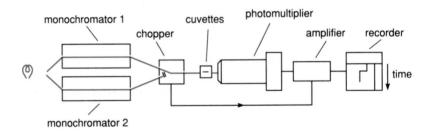

(c) Rapid kinetics: stopped flow in combination with dual wavelength

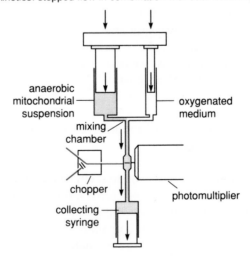

c oxidoreductase, or *bc*$_1$ complex) and complex IV (cytochrome *c* oxidase). Complex V is another name for the ATP synthase (Chapter 7).

The electron transfer activity of each complex is retained during this solubilization, and when complexes I, III or IV are reconstituted into artificial bilayer membranes, their ability to translocate protons is restored. Fractionation and reconstitution of the complexes has served a number of purposes: (1) the complexity of the intact mitochondrion is reduced; (2) it is possible to establish the minimum number of components which are required for function; (3) during the period in which the chemiosmotic theory was being tested, reconstitution proved one of the most persuasive techniques for eliminating the necessity of a direct chemical or structural link between the respiratory chain and the ATP synthase. For example, it proved possible to 'reconstitute' ATP synthesis by combining such disparate complexes as the bovine-heart mitochondrial ATP synthase and the light-driven proton-pump from halobacteria (Section 6.5) in a single bilayer.

There appear to be two moles of complex IV per mole of complex III, while complex I and complex II are present at a substantially lower stoichiometry. The isolated complexes readily reassemble. For example, in the presence of phospholipid and UQ$_{10}$, complex I and complex III reassemble spontaneously to reconstitute NADH–cytochrome *c* oxidoreductase activity. The technical problems surrounding reconstitution are considerable. First, the complex must be incorporated into a bilayer in a way which retains catalytic activity. Secondly, allowance has to be made for the possibility of a random orientation of the reconstituted proton pumps which would prevent the detection of net transport (Fig. 1.8). Incorporation into vesicles is normally accomplished by

Figure 5.2 Spectroscopic techniques for the study of the respiratory chain.

(a) The split-beam spectrophotometer uses a single monochromator, the output from which is directed alternately (by means of a chopper oscillating at about 300 Hz) into reference and sample cuvettes. A single large photomultiplier is used and the alternating signal is amplified and decoded so that the output from the amplifier is proportional to the difference in absorption between the two cuvettes. If the monochromator wavelength is scanned a difference spectrum is obtained. The split beam is therefore used to plot difference spectra which do not change with time. (b) The dual wavelength spectrophotometer uses two monochromators, one of which is set at a wavelength optimal for the change in absorbance of the species under study and one set for a nearby isosbestic wavelength at which no change is expected. Light from the two wavelengths is sent alternately through a single cuvette. The output plots the difference in absorbance at the two wavelengths as a function of time, and is therefore used to follow the kinetics or steady-state changes in the absorbance of a given spectral component, particularly with turbid suspensions. (c) To improve the time resolution of the dual wavelength spectrophotometer a rapidmixing device can be added. The syringes are driven at a constant speed and the 'age' of the mixture will depend on the length of tubing between the mixing chamber and the cuvette. When the flow is stopped, the transient will decay and this can be followed.

suspending the complex in cholate together with phospholipid, and then slowly dialysing away the detergent. Vesicles with the complex embedded in the membrane form spontaneously. Alternatively the complex can be sonicated together with the phospholipid. Such reconstitution experiments provide an approach to establishing the order of the components in the chain.

5.2.2 Methods of detection of redox centres

(a) Cytochromes

The cytochromes were the first components to be detected, due to their distinctive, redox-sensitive, visible spectra. An individual cytochrome exhibits one major absorption band in its oxidized form, while most cytochromes show three absorption bands when reduced. Absolute spectra, however, are of limited use when studying cytochromes in intact mitochondria or bacteria, owing to the high non-specific absorption and light-scattering of the organelles. For this reason, cytochrome spectra are studied using a sensitive differential, or split-beam, spectroscopy in which light from a wavelength scan is divided between two cuvettes containing incubations of mitochondria identical in all respects except that an addition is made to one cuvette to create a differential reduction of the cytochromes (Fig. 5.2). The output from the reference cuvette is then automatically subtracted from that of the sample cuvette, to eliminate nonspecific absorption. Figure 5.3 shows the reduced, oxidized, and reduced minus oxidized spectra for isolated cyt c, together with the complex reduced minus oxidized difference spectra obtained with SMPs, in which the peaks of all the cytochromes are superimposed.

The individual cytochrome may most readily be resolved on the basis of their α-absorption bands in the 550–610-nm region. The sharpness of the spectral bands can be enhanced by running spectra at liquid N_2 temperatures (77 °K), due to a decrease in line broadening resulting from molecular motion and to an increased effective light path through the sample resulting from multiple internal reflections from the ice crystals (Fig. 5.3).

Room-temperature difference spectroscopy can only clearly distinguish single a-, b- and c-type cytochromes. However, each is now known to comprise two spectrally distinct components. The a-type cytochromes can be resolved into a and a_3 in the presence of CO, which combines specifically with a_3. a and a_3 are chemically identical but are in different environments. The b-cytochromes consist of two components with different E_m values (high b_H and low b_L). These respond differently when a Δp is established across the membrane (Section 3.6.2). It is now clear (Section 5.8) that the two components reflect the presence on one polypeptide chain of two b-type haems; the different local environments provided by the polypeptide chain account for the differences in spectral and redox properties.

The two c-type cytochromes, cyt c and cyt c_1, can be resolved spectrally at low temperatures. Cyt c_1 is an integral protein within complex III (Section

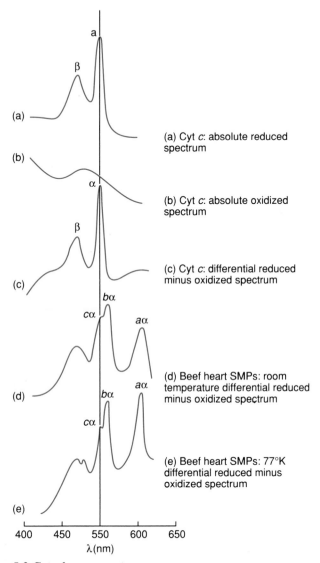

Figure 5.3 Cytochrome spectra.
The absolute oxidized (a) and reduced (b) spectra were obtained with purified cyt c in a split-beam spectrophotometer with water in the reference
cuvette. The reduced minus oxidized spectrum (c) was obtained with
reduced cyt c in one cuvette and oxidized cyt c in the other. (d) shows the
reduced (with dithionite) minus oxidized (with ferricyanide) spectrum from
beef heart SMPs. In (e) the scan was repeated at 77 °K, note the greater
sharpness of the α-bands

5.8), while cyt *c* is a peripheral protein on the P-face of the membrane and links complex III with cytochrome *c* oxidase.

(b) *Fe/S centres*

While their distinctive visible spectra aided the early identification and investigation of the cytochromes, the other major class of electron carriers, the iron–sulphur (Fe/S) proteins (Fig. 5.4), have ill-defined visible spectra but characteristic electron spin resonance spectra (ESR or EPR); see Fig. 5.5 . The unpaired electron, which may be present in either the oxidized or reduced form of different Fe/S proteins, produces the ESR signal. Each Fe/S group which can be detected by ESR is termed a centre or cluster. A single polypeptide may contain more than one centre. It is not clear how many centres are present in the mitochondrial respiratory chain; complex I may have up to 7 (Section 5.6).

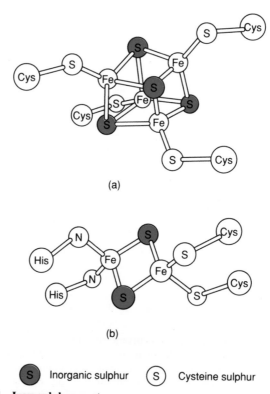

(a)

(b)

Ⓢ Inorganic sulphur   Ⓢ Cysteine sulphur

Figure 5.4   Iron-sulphur centres.
(a) A centre with four Fe and four acid-labile sulphurs is shown. On treatment with acid these sulphurs (blue) are liberated as H_2S. Although there are four Fe atoms, the entire centre undergoes only a one-electron oxido-reduction. (b) The deduced structure of the 2 Fe/S centre in complex III with two histidine ligands is shown; other 2Fe/S structures will have four cysteine ligands to the Fe.

(a) Apparatus

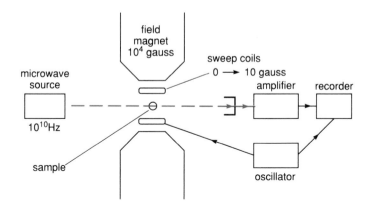

(b) Ideal absorption peak   (c) Differential   (d) Actual differential spectrum

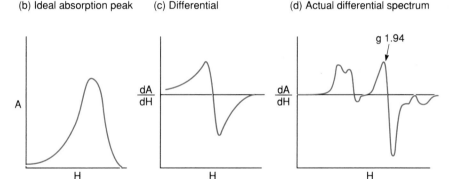

Figure 5.5   Electron spin resonance and the detection of Fe/S centres.
(a) Apparatus: a microwave source produces monochromatic radiation at
about 10^9 Hz, 30 cm. Unpaired electrons in the sample absorb the radiation when a magnetic field is applied, the precise value of the field
required for absorption depending on the molecular environment of the
electron, according to the formula:

$$h\nu = g\beta H$$

where h is Planck's constant, ν the frequency of the radiation, β a constant, the Bohr magneton, H the applied magnetic field, and g the spectroscopic constant which is diagnostic of the species. The spectrum is
obtained by keeping the microwave frequency constant and varying the
magnetic field, trace (b). In practice, a differential spectrum, trace (c), is
obtained by superimposing upon the steadily increasing field a very rapid
modulation of small amplitude obtained with auxiliary sweep coils. The
change in microwave absorption across each of these sweeps enables the
differential to be obtained. Spectra of energy-transducing membranes are
complex, trace (d); g values are obtained from either the peaks, the troughs
or the points of inflexion of the trace. Samples must be frozen and generally present at a high protein concentration (10–50 mg protein ml^{-1}).

Fe/S proteins contain Fe atoms covalently bound to the apoprotein via a cysteine sulphur and bound to other Fe atoms via acid-labile sulphur bridges (Fig. 5.4). Fe/S centres may contain two or four Fe atoms, even though each centre only acts as a one-electron carrier. Fe/S proteins are widely distributed among energy-transducing electron-transfer chains and can have widely different $E_{m,7}$ values, from as low as -530 mV for chloroplast ferredoxin (Section 6.4) to $+360$ mV for a bacterial HiPIP ('high-potential iron–sulphur protein').

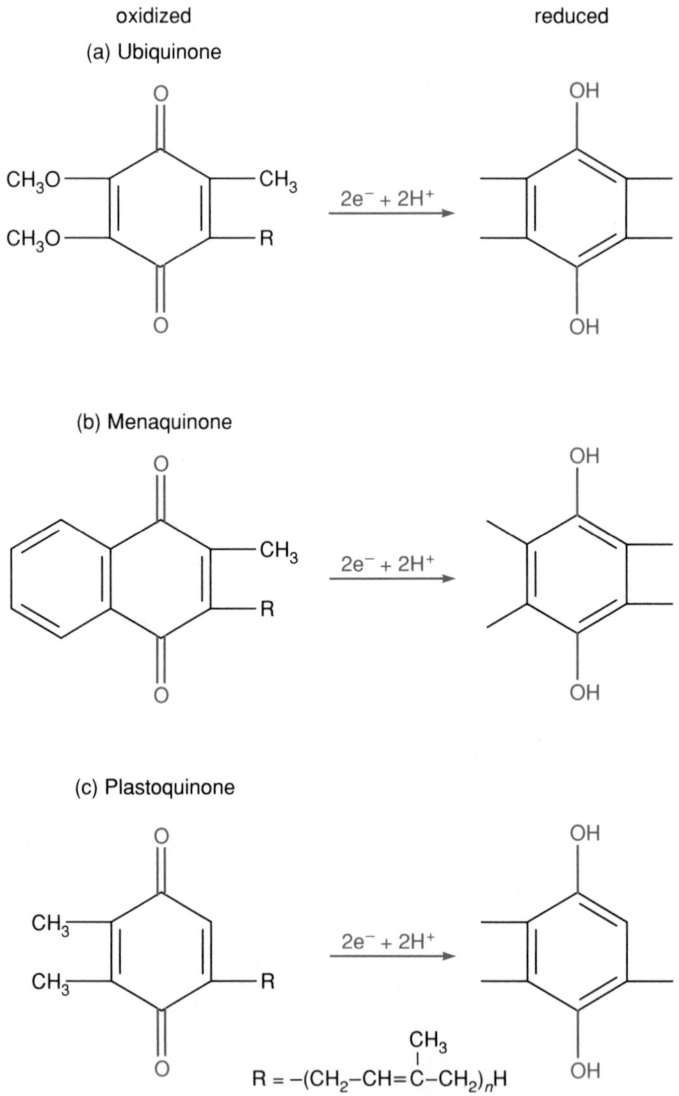

Figure 5.6   **Ubiquinone and related redox carriers.**

(c) *Quinones*

The 50-carbon hydrocarbon side chain of ubiquinone renders UQ_{10} highly hydrophobic (Fig. 5.6). UQ undergoes a $2H^+ + 2e^-$ reduction to form UQH_2 (ubiquinol), although a partially reduced free radical form UQH (ubisemi-quinone) plays a role in complex III, and the non-protonated, fully reduced form UQ^{2-} has a transient existence in both the photosynthetic reaction centre and the cyt bc_1 complex.

The radical form can be detected by its ESR spectrum or a characteristic absorption band in the visible region, but ubiquinone and ubiquinol are more difficult to detect because in common with proteins they absorb around 280 nm, although the absorbance of the oxidized and reduced forms differ.

The role of UQ in the respiratory chain was a matter of some controversy. The simplest postulate for the role of UQ is as a mobile redox carrier linking complexes I and II with complex III, although the 'Q-cycle' of electron transfer in complex III proposes a more integral role (Section 5.8). While UQ_{10} is the physiological mediator, its hydrophobic nature makes it difficult to use experimentally, and ubiquinones with shorter side chains, and consequently greater water solubility, are usually employed.

Some anaerobic respiratory chains employ menaquinone in place of UQ (Section 5.13.2) while in the chloroplast the corresponding redox carrier (Section 6.4) is plastoquinone (Fig. 5.6).

5.3 THE SEQUENCE OF REDOX CARRIERS IN THE RESPIRATORY CHAIN

The sequence of electron carriers in the mitochondrial respiratory chain (Fig. 5.1) was largely established by the early 1960s as a result of the application of oxygen-electrode (Fig. 4.11) and spectroscopic techniques. This work was greatly facilitated by the ability to feed in and extract electrons at a number of locations along the respiratory chain, corresponding to the junctions between the respiratory complexes. Thus NADH reduces complex I, succinate reduces complex II, and TMPD reduces cytochrome oxidase via cyt *c*. In this last case, ascorbate is usually added as the reductant to regenerate TMPD from its oxidized form known as Wurster's blue (WB). Ferricyanide (hexa-cyanoferrate(III)) is a non-specific, but impermeant, electron acceptor and can be used not only to dissect out regions of the respiratory chain but also to provide information on the orientation of the components within the membrane.

It should be emphasized that it is now clear that the electron carriers do not operate in a simple linear sequence, but that electrons may divide between carriers in parallel (as happens with complex III) or may be temporarily stored on components to enable a multi-electron reduction to occur (as occurs in the $4 e^-$ reduction of O_2 to H_2O in cytochrome oxidase).

The discovery of specific electron-transfer inhibitors enabled the relative positions of sites of electron entry and inhibitor action to be determined

(Fig. 4.12). Armed with this information, it was possible to proceed to a spectral analysis of the location of each redox carrier relative to these sites.

An independent approach to the ordering of the redox components came with the development of techniques for studying their kinetics of oxidation following the addition of oxygen to an anaerobic suspension (Fig. 5.2). The sequence in which the components become oxidized can reflect their proximity to the terminal oxidase and also whether they are kinetically competent to function in the main pathway of electron transfer. The rapidity of the oxidations observed under these conditions requires the use of stopped-flow techniques (Fig. 5.2).

The carriers in the respiratory chain must be ordered in such a way that their operating redox potentials E_h (Section 3.3.3) form a sequence from NADH to O_2. E_h is determined from the midpoint potential, E_m, and the extent of reduction (Eq. 3.20). Although the extent of reduction of a component in the respiratory chain can be measured spectroscopically, indirect methods are needed to measure the mid-point potential *in situ*. It should be noted that the mid-point potential of a component in the respiratory chain is usually different from that of the purified, solubilized component.

5.4 THE MECHANISM OF ELECTRON TRANSFER

Review   Moser *et al.* (1992)

How do electrons pass from one redox centre to another? In most circumstances the redox centres are not physically in contact but are separated by several nanometres. Thus, using terminology from inorganic chemistry, it is 'outer sphere' mechanisms rather than 'inner sphere' mechanisms that appear to be applicable. The most detailed information is available for mitochondrial cyt *c*.

The structures of cyt *c* from several mitochondrial sources have been elucidated. The haem sits in a largely hydrophobic crevice with only one edge (i.e. part of one porphyrin ring) exposed to solvent (Fig. 5.7). As we shall see (Sections 5.8 and 5.9) in mitochondria cyt *c* is reduced by the bc_1 complex and oxidized by cytochrome *c* oxidase. A fundamental question is whether the electron enters and leaves by the same route. Evidence that implicates a single route is that chemical modification of a group of lysine residues (especially 13, 86 and 87) found near the exposed haem edge in all cyt *c* molecules inhibits electron transfer both to and from the haem. These lysines appear therefore to be involved in electron transfer to the haem; interestingly, they are protected from chemical modification by complex formation with the physiological redox partner.

Clearly, if a single patch of the surface of the cytochrome is responsible for the protein–protein interaction with the redox partner, it follows that after reduction by the bc_1 complex, the cyt *c* must dissociate before forming a productive complex with cytochrome *c* oxidase for the oxidation reaction to occur. This is in accord with the current view of the electron-transport system

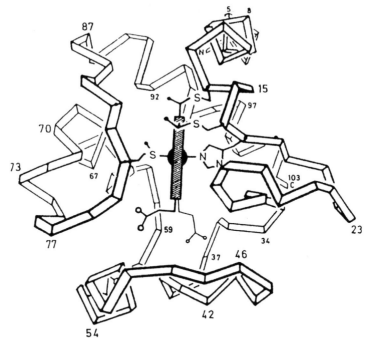

Figure 5.7   An outline of the structure of mitochondrial cytochrome *c*.
The haem and the ligands are in the centre and the haem is covalently (in a post-translational reaction) linked to the polypeptide. Residue numbers start from the N-terminus. Reproduced with permission from *The Biological Chemistry of the Elements* by J.J.R. Fraústo da Silva and R.J.P. Williams, Oxford University Press, Oxford, which should be consulted for further information about metal centres in redox proteins.

in which the cyt bc_1 and oxidase complexes are thought to diffuse relatively slowly whilst the peripheral protein, cyt *c*, undergoes more rapid lateral diffusion along the surface. The other relatively mobile component in the respiratory chain is UQ/UQH_2 which is believed to accept and donate electrons through respective collisions with dehydrogenases and the bc_1 complex.

It is most unlikely that there is direct interaction between the haem of cyt *c* and that of an acceptor or donor complex. Indeed it is clear from known structures, e.g. the tetra-haem subunit of the reaction centre of *Rhodopseudomonas viridis* (Section 6.2.3), that the electron is able to travel through a protein environment over considerable distances (up to approximately 2 nm). It is not known to what extent electron-transfer processes in general are facilitated by side chains of amino acids believed to lie close to the pathway of electron flux. At one time it was thought that certain conserved aromatic amino acids were vital. Thus on cyt *c*, Phe-82 (in the horse heart protein) lies on a putative route from the haem to the surface of the protein. However, site-specific mutagenesis has eliminated an obligatory role for this residue.

It has nevertheless been possible to establish (Cramer and Knaff, 1989; Moore and Pettigrew, 1990) that long-range electron transfer through proteins depends on several factors:

- the driving force, i.e. the redox potential difference, ΔE_h
- the relative orientation of the donor and acceptor in the reaction
- the distance separating the donor and acceptor
- the relaxation energy associated with the conformational change that accompanies the change of redox state.

The intervening medium, in the sense of amino acid side chains, is only definitely known to participate directly in electron transfer when a very electropositive redox centre is involved (Sections 5.12 and 6.4.2).

5.4.1 Redox potentiometry

The technique of redox potentiometry (Fig. 5.8) combines dual wavelength spectroscopy with redox potential determinations. As with redox potentiometry of most biological couples, it is necessary to add a low concentration of an intermediate redox couple in order to speed the process of equilibrium between the platinum electrode and the primary couple. As a secondary mediator will only function effectively in the region of its mid-point potential (so that there are appreciable concentrations of both its oxidized and reduced forms), a set of mediators is required to cover the whole span of the respiratory chain, with mid-point potentials spaced at intervals of about 100 mV. Mediators are usually employed at concentrations of 10^{-6} to 10^{-4} M. Many mediators are autoxidizable, and the incubation has to be maintained anaerobic for this reason and also to prevent a net flux through the respiratory chain from upsetting the equilibrium.

A second requirement for membrane-bound systems is that the mediators must be able to permeate the membrane in order to equilibrate with all the components. This introduces a considerable complication if the mitochondria are studied in the presence of a $\Delta\psi$ (i.e. in the presence of ATP since the incubation must be anaerobic) since $\Delta\psi$ or ΔpH will affect the distribution of the oxidized and reduced forms of the mediators across the membrane, and the oxidized/reduced ratio of the mediator at the site of the component will differ from that at the platinum electrode. Interpretation of ATP-dependent effects are thus complicated, as the possibility of this artifact must be distinguished from any genuine effect due to redistribution of electrons across the membrane on induction of a membrane potential. The simplest redox potentiometry is therefore performed with mitochondria or SMPs at zero Δp.

The practical determination of the E_m of a respiratory chain component (Fig. 5.8) involves incubating mitochondria anaerobically in the presence of the secondary mediators. The state of reduction of the relevant component is monitored by dual wavelength spectrophotometry (Fig. 5.2), while the ambient redox potential is monitored by a platinum or gold electrode. The electrode allows the secondary mediators and the respiratory chain components all to

(a) Apparatus

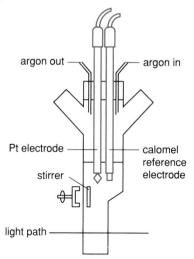

(b) Redox difference spectra succinate–cytochrome c reductase, Complex(II+III)

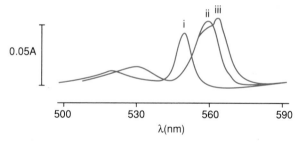

Figure 5.8   Redox potentiometry of respiratory chain components.
(a) Apparatus for the simultaneous determination of redox potential
and absorbance. (b) Difference spectra obtained with a suspension of
succinate–cytochrome c reductase (i.e. complexes II + III). The complex,
held in solution by a low concentration of detergent, was added to an
anaerobic incubation containing redox mediators. The ambient redox
potential was varied by the addition of ferricyanide. (i) Reference scan
(baseline) at $+280$ mV (all cytochromes oxidized), second scan at
$+145$ mV (cyt c_1 now reduced). (ii) Baseline at $+145$ mV (cyt c_1 reduced),
second scan at 10 mV (cyt b_L additionally reduced). (iii) Baseline at
-10 mV (c_1 and b_L reduced), second scan at -100 mV (b_H additionally
reduced). Data adapted from Dutton (1978).

equilibrate to the same E_h. This potential can then be made more electronegative (by the addition of ascorbate, NADH, or dithionite) or more electropositive by the addition of ferricyanide, and E_h and the degree of reduction
of the component are monitored simultaneously. In this way a redox titration
for the component can be established.

Considerable information can be gathered from such a titration. Besides E_h itself, the slope of $\log_{10}$ ([ox]/[red]) establishes whether the component is a one-electron carrier (60 mV per decade) or a two-electron carrier (30 mV per decade), (Table 3.2). By repeating the titration at different pH values it can be seen whether the mid-point potential is pH dependent, implying that the component is a $(H^+ + e^-)$ carrier. Finally, the technique frequently allows the resolution of a single spectral peak into two or more components based on differences in E_m. In this case the basic Nernst plot (Fig. 3.3) is distorted, being the sum of two plots with differing E_m values which can then be resolved. One of the most interesting findings with this technique was that cyt b in complex III can be resolved into two components (Section 5.8). Redox potentiometry can also be employed for the Fe/S proteins, in which case the redox state of the components is monitored by ESR.

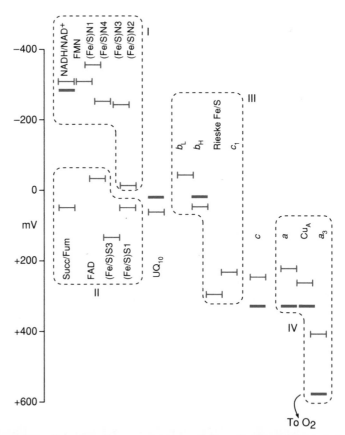

Figure 5.9 E_m **values for components of the mitochondrial respiratory chain and** E_h **values for mitochondria respiring in state 4.**
Values are consensus values for mammalian mitochondria. (—) $E_{m,7}$ values obtained with de-energized mitochondria, (—) $E_{h,7}$ values for mitochondria in state 4.

5.4.2 E_h values for respiratory chain components fall into isopotential groups separated by regions where redox potential is coupled to proton translocation

The mid-point potentials of some of the identifiable components of the respiratory chain are depicted in Fig. 5.9. Once the E_m values have been established for non-respiring mitochondria, an E_h for a component can be assigned to any component in respiring mitochondria simply by determining the degree of reduction. The results for mitochondria respiring in state 4 are shown. The oxido-reduction components fall into four equipotential groups, the gaps between which correspond to the regions where proton translocation occurs. The drop in E_h of the electrons across these gaps is conserved in Δp.

5.5 PROTON TRANSLOCATION BY THE RESPIRATORY CHAIN; 'LOOPS' OR 'CONFORMATIONAL PUMPS'?

For many years there was a controversy among convinced adherents to the chemiosmotic theory as to the model which most accurately represented the mechanism of respiratory chain proton translocation. The origins of the theory as developed by Peter Mitchell predated our present knowledge on the conformational, flexibility of proteins and proposed that the proteins of the respiratory chain provided a rigid and largely passive scaffolding for the redox carriers to carry out a vectorial (directional) translocation of protons (Fig. 4.9). The primary event was proposed to be electron transfer along a sequence of redox carriers arranged in the membrane such that transfer of electrons from a $(H^+ + e^-)$ carrier to a pure electron carrier took place with the release of protons on the P-side of the membrane, whereas electron transfer from a pure carrier to a $(H^+ + e^-)$ carrier took place with the uptake of protons from the N-side (Fig. 4.9). Protonation following addition of an electron to a redox carrier is common, and reflects an increased proton affinity (raised pK_A) of the reduced redox species (see, for example, Section 3.3). This type of mechanism has a limiting stoichiometry of $1\,H^+$ per electron per 'loop'.

It is sometimes postulated that the 'loop' hypothesis requires that the loops should span the membrane and that the appropriate redox carriers should be detectable on the P-face of the membrane. However, all that is required for a functional loop is that there should be a means for releasing the proton on the P-face. Thus, all the redox carriers could be located close to the N-face of the membrane, proton translocation being catalysed by specific proton channels analogous to the F_0 component of the ATP synthase (Section 7.4).

In a purely *conformational pump* model a redox carrier anchored within a more flexible protein is proposed to undergo redox-induced changes in pK_A, the directionality of proton transport being assured by co-ordinate conformational changes which make the redox site alternately accessible from either side of the membrane (Fig. 5.10). The reversibly protonated site in this model need not necessarily be limited to a conventional redox centre – it is equally possible

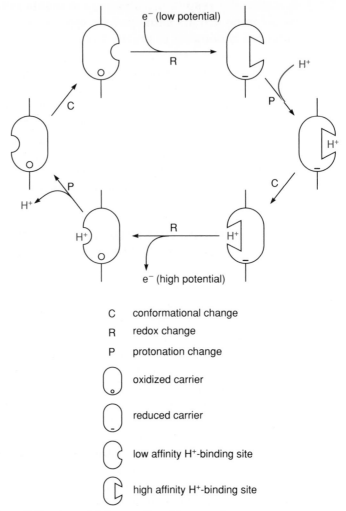

C	conformational change
R	redox change
P	protonation change
	oxidized carrier
	reduced carrier
	low affinity H⁺-binding site
	high affinity H⁺-binding site

Figure 5.10   A model for a redox-driven proton pump.
A hypothetical model is shown for a redox pump with a stoichiometry of
$1H^+/e^-$. The pump can exist in eight states (the combinations resulting
from two conformations, two redox states and two protonation states). In
order to produce a vectorial proton pump, only certain transitions are
allowed.

that a redox-induced conformational change alters the pK_A of an amino acid
side chain.

A conformational redox pump must co-ordinate a redox change, a conformational change, and a protonation change. Therefore even the simplest
model has 2^3 (i.e. 8) hypothetical states interconnected by a cubic array of
transitions. Vectorial proton translocation would be imposed by restricting the
number of permitted transitions as shown in Fig. 5.10.

A rather unconstructive debate persisted in the research literature for some
years as to whether the loop or conformational models were most accurate. It

is now apparent that both mechanisms coexist in nature: of the two ion pumps which have been characterized at atomic or near-atomic resolution − the bacterial photosynthetic reaction centre (Section 6.2.2) and the proton-pumping bacteriorhodopsin (Section 6.5) − the former functions as a sophisticated redox loop, with a relatively rigid array of electron carriers traversing the membrane, whereas the latter is a classic example of a conformational pump co-ordinating conformational and pK_A changes.

In the respiratory chain the complex for which there is the greatest consensus, the bc_1 complex, appears to function as a variant of a redox loop, with electrons shuttling between separate sites of proton uptake and release (Section 5.8). On the other hand ATP-driven ion pumps, which lack redox centres, can be rationalized as conformational pumps, transporting not only H^+, but also Na^+, Ca^{2+} and K^+, across the appropriate membranes.

5.6 COMPLEX I (NADH–UQ OXIDOREDUCTASE)

Reviews   Weiss *et al.* 1991, Walker, 1992

There are two reasons why NADH–UQ oxidoreductase (complex I) is the least well understood component of the mitochondrial electron-transport chain. In the first instance it is large (about the same size as the large subunit of a ribosome), has more than thirty polypeptides, and, unlike complexes III and IV (see below), no simpler but functionally similar counterpart has yet been isolated from a bacterium. Secondly the redox centres, apart from one molecule of the flavin FMN, are Fe/S centres, which cannot be studied by optical spectroscopy but instead require low-temperature ESR (Section 5.2.2b) to assess their redox state. What is certain about complex I is that it catalyses the transfer of two electrons from NADH to ubiquinone in a reaction that is associated with proton translocation across the membrane and is inhibited by rotenone and piericidin A. Current evidence suggests that the proton-translocation stoichiometry is $4H^+/2e^-$.

There may be as many as seven Fe/S centres in the complex. Four of these, called N-1, N-2, N-3 and N-4, are well characterized, both in bovine heart and *Neurospora crassa* mitochondria. N-1 is a 2Fe/2S centre whilst the other three are 4Fe/4S centres (Section 5.2.2b). The N-2 centre has a $E_{m,7}$ in the range -20 to -160 mV (the exact value depending on the source of the complex) whilst the other three are more negative; for example, in *N. crassa* $E_{m,7}$ values of -330 mV (N-1), -230 mV (N-3) and -300 mV (N-4) have been reported. Using the $E_{m,7}$ values as a guide (but recalling that it is the values of E_h that are relevant) a likely pathway of electron flow may be NADH → FMN → N-1 → N-4 → N-3 → N-2 → UQ. Note, however, that the precedent of the bc_1 complex (Section 5.8) allows for a non-linear electron flow pathway.

Complex I can be split by treatment with chaotropic ions (these, e.g. Br^-, disrupt water structure) into three parts. One of these is water-soluble, catalyses transfer of electrons from NADH to non-physiological acceptors such as ferricyanide, contains not only the NADH binding site but also two Fe/S centres (2Fe/2S and 4Fe/4S) plus FMN and is known as the flavoprotein (FP) fraction. It contains three polypetides of M_r 51 000, 24 000 and 10 000. A second fraction comprises approximately six polypeptides (one of which has a M_r of 75 000), contains at least three Fe/S centres and is accordingly known as the iron–sulphur protein (IP) fraction. The third fraction is the residual material, is known as the hydrophobic (HP) fraction because of its insolubility in water and also contains Fe/S centres.

Complete DNA sequences, and from there amino acid sequences, have now been obtained for many of the polypeptides of the enzymes from both bovine heart and *N. crassa*. Some striking findings have emerged. The sequence data for the 75-kDa peptide from IP and the 51-kDa and 24-kDa peptides from FP, none of which appear to have hydrophobic transmembrane α-helices, indicates that regions of these polypeptides have considerable homology to sequences within two subunits, α and γ, of a water-soluble H_2–NAD^+ oxidoreductase found in the bacterium *Alcaligenes eutrophus* (note, this is distinct from the bacterial hydrogenase discussed in Section 5.13). A dimer of the α- and γ-subunits in this type of enzyme from a related bacterium contains FMN, Fe/S centres (probably two 4Fe/S and one 2Fe/2S clusters) and has NADH–ferricyanide oxidoreductase activity in common with the FP segment of NADH dehydrogenase. The *A. eutrophus* α-subunit has considerable homology to sizeable stretches of both the 51-kDa and the 24-kDa subunits of FP whilst its γ-subunit resembles the 75-kDa subunit of IP. These comparisons and other features strongly suggest that the 51-kDa subunit contains the site at which electrons are transferred to FMN from NADH.

Both the 51-kDa subunit of FP and the α-chain of the *A. eutrophus* enzyme contain a Cys-X-X-Cys-X-X-Cys motif, strongly indicating the presence of an Fe/S centre. This is likely to be a 4Fe/4S centre and that on the 51-kDa sub-unit may be centre N-3. The homologous regions of the γ-subunit of the *A. eutrophus* enzyme and the 75-kDa subunit of IP contain ligands for a second, and possibly a third, 4Fe/4S centre. Finally, homologous regions of the 24-kDa peptide of FP from several sources including the bacterium *Paracoccus denitrificans* and the α-subunit of *A. eutrophus* H_2–NAD^+ oxidoreductase, contain Cys residues which could contribute to a fourth Fe/S centre (probably 2Fe/2S type). In summary, the relationship between the two homologous complexes has established the role of the 75-, 51- and 24-kDa subunits and their redox centres in the early steps of the NADH dehydrogenase reaction.

As expected, the polypeptides in the hydrophobic fraction (HP) of complex I contain many potential transmembrane α-helices. Seven of these peptides are coded for by what were formerly termed unidentified reading frames (URFs), but now termed *ndh* genes, in mitochondrial DNA. HP is expected to contain the N-2 Fe/S centre which by virtue of its highest (i.e least negative) $E_{m,7}$ is probably the donor of electrons to UQ. Three or four of the HP polypeptides have some similarity to the formate–H_2 lyase system of *E. coli* and one of these (*ndh-1*) also has similarity to part of a quinone-linked, piericidin-A-sensitive glucose dehydrogenase from *Acinetobacter calcoaceticus* and *E. coli*. The similarity of the latter enzyme to parts of the NADH dehydrogenase and the view that piericidin-A may be able to compete at a ubiquinone binding site

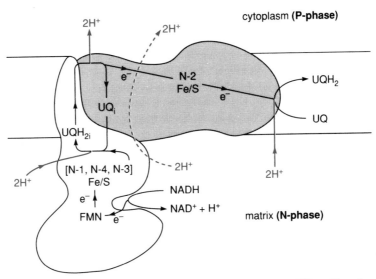

Figure 5.11   A structure for the mitochondrial NADH–UQ oxidoreductase (complex I).

This is based on evidence from electron microscopic studies, sequence data and the manner in which complex I splits into different modules (Weiss *et al.*, 1991). UQ_i is a putative internal ubiquinone. The proposed $4H^+/2e$ translocation stoichiometry is shown; the mechanism is unknown. The shaded region is the hydrophobic part of the enzyme.

has suggested candidates amongst the HP subunits of the NADH dehydrogenase which might be involved in UQ reduction. There is also tentative evidence for an internal UQ molecule, involved in H^+ translocation but not exchangeable with the bulk UQ pool, similar to that found in photosynthetic reaction centres (Section 6.2.2).

A tentative structure, based on the above information and consistent with low resolution electron microscope images, is shown in Fig. 5.11. It appears that the enzyme is 'modular' with electrons passing sequentially from module to module. The reason for such subunit complexity in NADH dehydrogenase is not clear. It will be important to discover how many subunits are required in bacterial NADH dehydrogenases, e.g. from *P. denitrificans*, that have mitochondrial-like characteristics. One possibility is that mitochondrial NADH dehydrogenase has other, unrecognized functions; in this connection one of the subunits has been shown to have some characteristics, including attached pantothenic acid, of an acyl carrier protein.

5.7 COMPLEX II (SUCCINATE DEHYDROGENASE); ELECTRON-TRANSFERRING FLAVOPROTEIN AND α-GLYCEROPHOSPHATE DEHYDROGENASE

In addition to complex I, three other redox pathways feed electrons to UQ_{10}: complex II, which transfers electrons from succinate; electron-transferring flavoprotein, supplying electrons from the flavoprotein-linked step of fatty acid β-oxidation; and *sn*-glycerophosphate dehydrogenase. The first two are located on the matrix face of the membrane and the last is on the outer face. All three are flavoproteins transferring electrons from substrate couples with midpoint potentials close to 0 mV. As would be expected on thermodynamic grounds, none is proton-translocating.

Complex II consists of several polypeptides. The two largest constitute succinate dehydrogenase, the larger containing a covalently bound FAD and two Fe/S centres, while the other contains an additional Fe/S centre. Complex II contains a cyt *b* of unknown function which is associated with the smaller polypeptides, equimolar with FAD and distinct from the cyt *b* of complex III (Section 5.8).

5.8 UBIQUINONE AND COMPLEX III (*bc*$_1$ COMPLEX OR UQ–CYTOCHROME *c* OXIDOREDUCTASE)

Reviews Trumpower, 1990a, 1990b

The transfer of electrons from ubiquinol to cyt *c* and the associated proton translocation is not a simple matter. The reaction is catalysed by complex III of the respiratory chain which is also termed the cyt *bc*$_1$ complex, or more appropriately the ubiquinol–cytochrome *c* oxidoreductase. This complex is also found in many bacteria (Section 5.13) and is similar in many respects to the plastoquinol–plastocyanin oxidoreductase, or cyt *bf* complex, of thylakoids (Section 6.3).

Cyt *bc*$_1$ complexes from several sources contain three polypeptide chains that carry the redox groups: an iron–sulphur protein, often called the Rieske protein after its discoverer, cyt *c*$_1$ and cyt *b*. The Rieske protein contains a 2Fe/2S cluster attached to the polypeptide by chelation of one Fe to two cysteines and the other to two histidine residues. The amino acid sequences of Rieske proteins suggest that the polypeptide chain will fold as a globular structure incorporating the 2Fe/2S centre and extending into the aqueous layer beyond the bilayer on the cytoplasmic (or P) face of the membrane (in the case of the mitochondrion) and anchored to the membrane via a hydrophobic N-terminus.

Cytochrome *c*$_1$ has a similar conformation to the Rieske protein, except in this case it is the C-terminus which provides the hydrophobic anchor. The cyt *b* subunit binds two haem groups which are believed from spectroscopic evidence to have axial histidine ligands. Comparison of several sequences has allowed plausible models to be made showing how the polypeptides could fold into eight transmembrane α-chelices. Four conserved histidine residues within the helices have the appropriate topology to provide the ligands for the haems. One haem, with an $E_{m,7}$ of about -100 mV and thus known as b_L (formerly called b_{566} because of its α-band absorption maximum), is located towards the P (cytoplasmic) side of the mitochondrial membrane. The second haem, b_H, $E_{m,7}$ $+50$ mV, (formerly b_{560} because of its α-band at approximately 560 nm) is predicted to be on the N (matrix) side of the membrane. There are additional polypeptides in most ubiquinol–cytochrome oxidoreductases, such as the mitochondrial bc_1 complex, but there is no evidence that they play any role in either electron transport or proton translocation and thus they are not considered further here. This view is supported by the finding that the bacterium *Paracoccus denitrificans* contains no additional polypeptides in its complex and yet is able to couple electron flow to proton translocation.

The probable pathway of electron flow through the bc_1 complex is at first sight convoluted and we shall take some time to describe it in detail (Fig. 5.12). The discussion is for the mitochondrion, but the bacterial bc_1 complexes are closely similar.

5.8.1 Stage 1: UQH$_2$ oxidation at Q$_p$

A pool of ubiquinone and ubiquinol exists in the membrane in large molar excess over the other components of the respiratory chain. The mid-point potential, $E_{m,7}$, for the UQH$_2$/UQ couple (in aqueous solution) is $+60$ mV (Fig. 5.9), while the actual $E_{h,7}$, which takes into account the extent of reduction of the carrier, is close to 0 mV. A molecule of UQH$_2$ from the pool diffuses to a binding site Q$_p$ (also termed Q$_o$ or Q$_z$) close to the *cytoplasmic* (or P) face of the mitochondrial membrane and adjacent to the Rieske protein (Fig. 5.12a). What appears to happen is that the oxidation of UQH$_2$ to UQ takes place in two stages:

1. The first electron is transferred from UQH$_2$ to the Rieske protein (Fig. 5.12a), releasing two protons to the cytoplasm and leaving the free radical semiquinone anion species UQ$^{\bullet -}$ at the Q$_p$ site.
2. The second electron is transferred to the b_L haem, which is also close to the P-face.

This proposal raises a number of questions. Firstly, is it thermodynamically feasible? The $E_{m,7}$ for the UQH$_2$/UQ$^{\bullet -}$ couple is about $+280$ mV, close to that for the Rieske protein, and 220 mV more *positive* than the $E_{m,7}$ of the two-electron oxidation ($+60$ mV, see above). This means that the semiquinone is unstable and explains why the electron acceptor needs to be so electropositive. It also implies that the second stage of the oxidation will be energetically favourable as the semiquinone seeks to lose its second electron, and this is reflected in the E_m for the UQ$^{\bullet -}$/UQ couple, which at -160 mV is 220 mV more *negative* than that of the two-electron oxidation. This first one-electron oxidation step has thus generated one highly reducing electron. The semiquinone anion can, under certain conditions, be detected by ESR and is also an intermediate of the bacterial photosynthetic reaction centre, which has a quinone binding site capable of stabilizing the free radical anion (Section 6.2.2b).

The electron received by the Rieske protein passes down the chain to cyt c_1, cyt c and cytochrome oxidase (Fig. 5.12a).

(a)

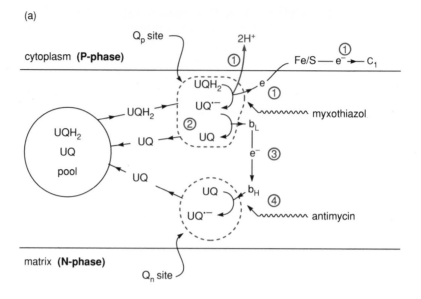

(b)

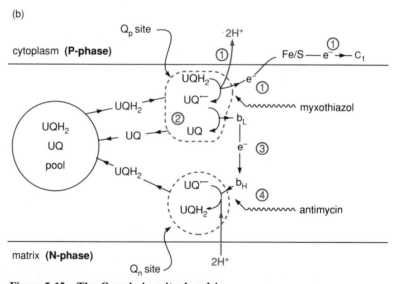

Figure 5.12   The Q-cycle in mitochondria.
(a) This illustrates the electron-transfer events that follow the oxidation of a ubiquinol at the P-side of the inner mitochondrial membrane under conditions in which the quinone binding site at the N-side is initially either vacant or occupied by a ubiquinone molecule. (b) This illustrates the electron-transfer events that follow the oxidation of a second ubiquinol at the P side of the membrane when the Q_n is occupied by a ubisemiquinone radical. Note that the Q_n site has also been termed the Q_i or Q_c site (c indicating the cytoplasmic side of the membrane in bacterial cytochrome bc_1 complexes) in various systems and the Q_p site is also known as the Q_o or Q_z site. The inhibitory sites of action of myxothiazol and antimycin are also shown by wavy arrows.

5.8.2 Stage 2: UQ reduction to UQ$\cdot^-$ at Q$_n$

The electron on b_L ($E_{m,7}$ -100 mV) now passes to the other haem, b_H ($E_{m,7}$ $+50$ mV in a de-energized mitochondrion) held on the same polypeptide. At first inspection this looks as though the electron is losing 150 mV of redox potential; however, the two haems are on different sides of the hydrophobic core of the membrane across which is a membrane potential of some 150 mV. (Fig. 5.12). This is known because the imposition of a membrane potential generated by ATP hydrolysis or by K$^+$ diffusion (Section 3.8) causes electrons to move from b_H on the N-side of the hydrophobic core to b_L on the P-side.

The relative position of the two haems means that in the normal operation of the complex the electron retains its original energy on passing from b_L to b_H since the drop to a more positive redox potential is compensated by the energetically unfavourable migration of the negative electron from the P-side to the N-side of the membrane. In an uncoupled mitochondrion energy would be dissipated at this step.

UQH$_2$ and UQ can migrate freely from one side of the hydrophobic core to the other regardless of $\Delta\psi$, since these hydrophobic carriers are uncharged. A second quinone binding site, Q$_n$, in the close vicinity of b_H is proposed which binds UQ and allows the transfer of the electron from the reduced b_H with the formation of the semiquinone anion UQ$\cdot^-$ (Fig. 5.12). At first glance this looks thermodynamically unlikely since the $E_{m,7}$ for the UQ$^-$/UQ couple in free solution is -160 mV while that for b_H is $+50$ mV. If, however, Q$_n$ were to bind the semiquinone much more firmly than UQ this would have the effect of shifting the $E_{m,7}$ to a more positive value, i.e. making the UQ more readily reducible. A ten-fold difference in the binding of the semiquinone relative to UQ shifts $E_{m,7}$ 60 mV more positive than if the reaction occurred in free solution. A 300-fold firmer binding of the semiquinone would thus make the $E_{m,7}$ 150 mV more positive.

We have not cheated the first law of thermodynamics here, since the energy required for the addition of a second electron to generate unbound UQH$_2$ (see below) is proportionately *increased*, i.e. the E_m for the couple UQH$_2$ $_{free}$/UQ$^-_{bound}$ is made proportionately more negative. This is confirmed by actual measurements of the two $E_{m,7}$ values, using ESR to detect the semiquinone. We will come across this concept of driving an apparently unfavourable reaction by making a product very firmly bound in Chapter 7, when we discuss the ATP synthase.

5.8.3 Stage 3: UQ$\cdot^-$ reduction to UQH$_2$ at Q$_n$

We now have a semiquinone firmly bound to Q$_n$. In the next part of the cycle (Fig. 5.12B) a second molecule of UQH$_2$ is oxidized at Q$_p$ in a repeat of stage 1 – one electron passing to cyt c_1 and the other via b_L to b_H. This second electron now completes the reduction of UQ$\cdot^-$ to UQH$_2$, the two protons required for this being taken up from the matrix (Fig. 5.12B). The UQH_2 returns to the bulk pool and the cycle is completed.

Q_n and Q_p are not equivalent in this model: only Q_n binds the semiquinone firmly – supported by its detection by ESR. At Q_p the redox potentials are widely separated and the semiquinone has only a transient existence.

5.8.4   The thermodynamics of the Q-cycle

The overall reaction catalysed by the bc_1 complex involves the *net* oxidation of one UQH_2 to UQ (two UQH_2 oxidized in stage 1 and one UQ reduced in stage 3), the reduction of two cyt c_1, the release of four protons at one side of the membrane and the uptake of two protons from the other. In the model we have discussed, the only *charge* transfer across the membrane is the movement of the two electrons between the haems which we placed on opposite sides of the hydrophobic barrier. In practice Q_p does appear to be close to the cytoplasmic face, such that the release of protons at Q_p does not significantly contribute to the displacement of charge across the membrane, whereas Q_n may be more deeply buried into the matrix side of the membrane such that the entry of protons from the matrix contributes partially to the charge movement.

So far we have an elegant model, but it may not be intuitively obvious why it can function as a proton *pump*, i.e. removing protons from the matrix at low electrochemical potential and releasing them in the cytoplasm at a Δp some 200 mV higher. To answer this we shall consider two conditions, where Δp is present purely as a membrane potential (approximating to the condition in the respiratory chain) and where Δp is present purely as a ΔpH (as would occur in a thylakoid membrane, where a closely analogous cycle probably operates; section 6.3), making the simplification that Q_n is close to the matrix face. In the first case the protons are present at equal concentrations on both sides of the membrane and the work which must be done is to push two electrons from the cytoplasmic to the matrix side of the membrane against a high membrane potential. As stated above this is energetically possible since the electrons are transferred from a negative (low) potential haem to a positive (high) potential haem. In the case of a pure ΔpH the electrons would flood from b_L to b_H since they have no $\Delta\psi$ to push against: this would drive UQH_2 oxidation at Q_p and UQ reduction at Q_n, enabling protons to be translocated against a high ΔpH.

One slightly puzzling feature at first sight is that for each two electrons, four protons appear in the cytoplasm when only two protons disappear from the matrix. This will be resolved when we discuss cytochrome c oxidase where $2H^+/2e^-$ appear in the cytoplasm and $4H^+/2e^-$ disappear from the matrix, restoring the balance.

5.8.5   Inhibitors of the Q-cycle

Antimycin and myxothiazol are two inhibitors of mitochondrial electron transport which act on the bc_1 complex. Antimycin acts at Q_n, preventing the formation of the relatively stable UQ$^{\cdot -}$. If, in the presence of this inhibitor, oxygen is added to an anaerobic suspension a *reduction* of the *b* cytochromes occurs following oxidation of UQH$_2$ at Q$_p$. This *oxidant-induced reduction* is consistent with the cycle but cannot be explained with a linear sequence of carriers. Myxothiazol inhibits events at Q$_p$ – it should be clear from inspection of Fig. 5.12 that oxidant-dependent reduction of the *b* cytochromes does not occur in the presence of this inhibitor.

5.9 CYTOCHOME *c* AND COMPLEX IV (CYTOCHROME *c* OXIDASE; FERROCYTOCHROME: O$_2$ OXIDOREDUCTASE)

Reviews Babcock and Wikström (1992), Capaldi (1990), Chan and Li (1990), Malmstrom (1990), Saraste (1990)

Complex III transfers electrons to cyt *c*. Cyt *c* is not isolated as a component of a complex, although it can bind stoichiometrically to cytochrome *c* oxidase. Cyt *c* is a peripheral protein located on the outer face of the mitochondrial membrane and may be readily solubilized from intact mitochondria. The detailed mechanism of cyt *c* was discussed in Section 5.4.

The final step in the electron-transport chain of mitochondria and certain species of respiratory bacteria operating under aerobic conditions is the sequential transfer of four electrons from the reduced cyt *c* pool to O$_2$, forming 2H$_2$O in a four-electron reaction catalysed by a cytochrome *c* oxidase:

$$O_2 + 4e^- + 4H^+ \rightarrow 2H_2O$$

In addition to the electron transfer, cytochrome *c* oxidase acts as a proton pump with a stoichiometry of about 1H$^+$/e$^-$. Because the mechanism of coupling between electron transfer and proton pumping is unclear the two aspects of the complex will be considered separately.

5.9.1 Electron transfer

The mammalian mitochondrial cytochrome *c* oxidase is also termed complex IV and contains two spectrally distinct, though chemically identical, haem *a* species – haem *a* and haem a_3 – together with at least two copper atoms known as Cu$_A$ and Cu$_B$ (the presence of a third is controversial at the time of writing) (Fig. 5.13). In some bacteria other types of cytochrome *c* oxidase may be present and equivalent *c*-type cytochromes may serve as electron donors (Section 5.13).

The structure of mitochondrial cytochrome *c* oxidase is exceedingly complicated, with more than ten subunits, the largest three of which are coded for by mitochondrial DNA. Bacterial cytochrome *c* oxidases, such as that from *P. denitrificans* (Section 5.13.1), in general only contain three subunits, corresponding to the three large subunits of the mitochondrial complex. Indeed, since only the two largest subunits are required for fully active electron transfer and since these possess the spectral characteristics of the complete oxidase,

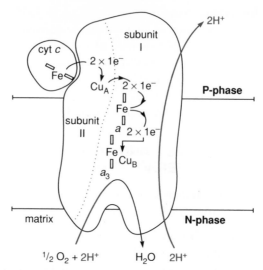

Figure 5.13   A tentative outline structure for cytochrome oxidase.
The general tooth-like shape has been deduced from electron microscopy.
The positions of the redox groups have been deduced from a range of
experiments, including model building of the likely folding patterns of the
two largest subunits that are thought to carry the main centres. It is usual
to show the site of oxygen reduction at the matrix side of the inner
mitochondrial membrane. This has not been established beyond doubt but
the protons required in the reduction are taken from the matrix. Thus the
charge translocation is achieved by a combination of the inward movement
of the electrons, and outward movement of protons to the bimetallic
centre, in addition to the proton pumping by the enzyme. See also Saraste
(1990) and Capaldi (1990).

it follows that the redox centres must be confined to these subunits. The location of the redox centres is not entirely certain, but analysis of the sequences
of the two largest subunits of the *P. denitrificans* oxidase suggests that the
haems and Cu_B are on the largest subunit and Cu_A on the smaller subunit.

Electrons from cyt *c* are initially transferred to Cu_A (Fig. 5.13). The other
redox components, Cu_B and haem a_3, are located close to each other and form
a binuclear centre which is probably located towards the N-side of the
membrane (Fig. 5.13).

The following reactions are believed to occur at the binuclear centre (Fig.
5.14):

(a) The Cu_B of the binuclear centre accepts one electron from the initial
acceptor (i.e. Cu_A or haem *a*) (O → H in Fig. 5.14)

(b) Receipt of a second electron by the centre (H → R) leads to binding of O_2
(R → A) and formation of a peroxy bridge between the Cu_B and the haem
a_3 (A → P). This can be regarded as the transfer of two electrons from the
binuclear centre to the O_2. Release of the oxygen species is stringently
avoided since this would lead to generation of the toxic superoxide radical
or peroxide.

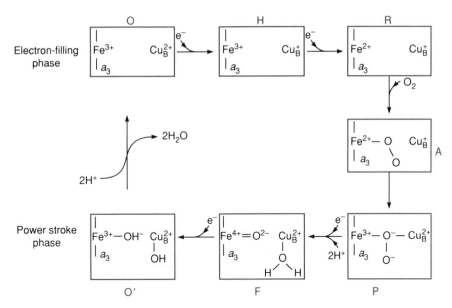

Figure 5.14   Tentative scheme for mechanism of oxygen reduction by cytochrome aa_3 oxidase.
For details see text. Adapted from Wikström and Babcock (1990).

(c) Protons are taken up from the N-phase and a third electron is supplied to the centre (P → F), resulting in breakage of the oxygen–oxygen bond, generating a ferryl species at the haem which is formally in the +4 oxidation state.

(d) In the final step of the catalytic cycle the arrival of a fourth electron (F → O′) results in the formation of a ferric hydroxide. Finally, further protons taken from the N-phase allow the co-ordinated hydroxyl groups to be lost as water and the binuclear centre to return to its unliganded oxidized state.

Although this scheme is very plausible, confirmation and further developments will probably require determination of a high-resolution structure as has been accomplished for the reaction centre (Section 6.2.2) and bacteriorhodopsin (Section 6.5). This will probably be realized by electron microscopy rather than X-ray crystallography; indeed a 1-nm resolution structure has already been obtained.

The oxygen binding site of cytochrome c oxidase also binds ligands such as CO, CN^- and N_3^- which explains why these species are inhibitors of respiration.

5.9.2 Proton pumping

The scheme of electron transfer discussed above has so far not involved the proton pumping which is known to occur. Electrons enter the complex from cyt c on the P-side at an $E_{h,7}$ of about $+290$ mV for mitochondria in state 4 and are ultimately transferred to the $\frac{1}{2}O_2/H_2O$ couple with an $E_{h,7}$ in air-saturated medium at about $+800$ mV. However, because the site of O_2 reduction is close to the N-side, the electrons have to cross the membrane against a $\Delta\psi$ of some 180 mV. This reduces the available energy just as the reverse process in the bc_1 complex increases the energy. The redox span $\Delta E_{h,7}$ is therefore over 300 mV ($800 - 290 - 180$ mV). Four electrons falling through this potential would be sufficient to translocate up to six protons across the membrane against a Δp of some 200 mV. However, unlike the remainder of the respiratory chain, cytochrome c oxidase is irreversible. The actual H^+/O stoichiometry is lower, probably $4H^+/4e^-$, reflecting this lack of reversibility.

The mechanism of proton pumping is unknown, although it is argued to be associated with steps $P \rightarrow F$ and $F \rightarrow O$ in Fig. 5.14 (Wikström and Babcock, 1990) and the protons may prove to pass close to the binuclear centre. Probably a conformational pump mechanism is involved. Subunit 3 has often been implicated on the basis of it being the site of the inhibitory action of DCCD, but site-directed mutagenesis studies have now established that this subunit is not an essential component of the proton-pumping mechanism (Haltia *et al.* 1991).

5.10 MITOCHONDRIAL ELECTRON TRANSPORT AND DISEASE

Reviews   Capaldi 1988, Cooper *et al.* 1991, Morgan-Hughes *et al.* 1990)

Mitochondrial myopathies are a group of human diseases which involve abnormalities in skeletal muscle mitochondria. Defects are found in each of the complexes I to IV which usually cause the rate of electron transport to be impaired. Increasingly, deficiencies in individual components of the complexes are being identified by antibody or gene probing procedures (Cooper *et al.* 1991). There is clearly tissue-specific expression of some of these polypeptides because other tissues may not be as severely affected as skeletal muscle although there are instances of brain mitochondrial dysfunction. Defects in the human respiratory chain lead to excess lactate production. In an attempt at therapy, membrane-permeable artificial redox carriers such as meandione have been administered to bypass a defect in complex III and hence to accelerate the rate of electron transport to oxygen. It is well known that inhibition of mitochondrial electron transport *in vitro* by myxothiazol or antimycin can be overcome by the addition of similar electron carriers. A different kind of defect is exemplified by Luft's syndrome. Here patients respire at relatively high rates but suffer muscular weakness and high temperatures. The contemporary explanation is that they have mitochondrial inner membranes that are unusually leaky to protons, thus causing shorting of the normal proton circuit (see Section 4.7.1).

Recently a protein in the inner mitochondrial membrane has been identified as playing a role in determining whether or not a cell takes the first step along a pathway leading to programmed cell death (Hockenbery *et al.*, 1990). This protein is coded for by the oncogene *bcl-2*. Its mechanism of action is unknown, but one possibility is that it might antagonize the activity of one or more nuclear-coded proteins that interfere with mitochondrial energy conservation or membrane permeability. It may also act in a fashion related to the hok toxin in bacteria (Grivell and Jacobs, 1991) which is responsible, by an action that both inhibits respiration and dissipates the protonmotive force, for the death of those cells that lose a plasmid at the point of cell division. If programmed death of mammalian cells were to be associated with the action of a respiratory toxin, then conceivably mutations in *bcl-2* could be connected with the prevention of the action of the toxin. Such an action of a toxin has an established parallel in plant mitochondria (Section 5.12).

5.11 THE NICOTINAMIDE NUCLEOTIDE TRANSHYDROGENASE

Review   Jackson, 1991

Although the midpoint potentials for the $NAD^+/NADH$ and $NADP^+/NADPH$ couples are the same (Table 3.2), the latter couple is considerably more reduced in the mitochondrial matrix. This disequilibrium is maintained by an energy-dependent transhydrogenase which catalyses the following reaction:

$$NADP^+ + NADH + nH_{out}^+ \rightleftharpoons NADPH + NADP^+ + nH_{in}^+$$

where n is probably 2. The observed mass-action ratio (Section 3.2) may exceed 500 and may be maintained either by respiration, in which case oligomycin is without effect, or by ATP hydrolysis, in which case oligomycin is inhibitory. This indicates that the transhydrogenase is dependent on Δp rather than ATP.

The mitochondrial transhydrogenase has one major polypeptide of 97 kDa which has been sequenced. The enzyme has 1043 residues with a 400-residue N-terminal cytoplasmic domain which binds NAD(H), a central hydrophobic region predicted to contain 14 transmembrane domains and a 200-residue C-terminal cytoplasmic domain which binds NADP(H). Susceptibility to tryptic cleavage is modified in the presence of NADPH (but not the other nucleotides), indicating that binding of NADPH causes a conformational change in the enzyme.

The transhydrogenase is an interesting exception to the rule that there should be a difference in mid-point potential across an energy-transducing step (Fig. 5.9).

5.12 ELECTRON TRANSPORT IN MITOCHONDRIA OF NON-MAMMALIAN CELLS

Reviews   Douce and Neuburger 1989, Moore and Siedow, 1991

The electron-transport systems of mitochondria from mammalian sources, especially liver and heart, are the most studied at the biochemical level, in part because large-scale preparation of mitochondria from these sources is relatively easy. This emphasis tends to obscure the significant differences that are found in mitochondria from other sources. As examples we give a brief overview of mitochondrial electron transport in plants, fungi and parasites.

A feature found in plant mitochondria that distinguishes them from their mammalian counterparts is the frequent presence of an electron-transport pathway from ubiquinol to oxygen that is independent of the cytochrome aa_3 and cytochrome bc_1 complexes. This pathway is characteristically inhibited by salicylhydroxamic acid (usually called SHAM) but by neither cyanide, antimycin nor myxothiazol (Fig. 5.15). It is regarded as an 'uncoupled' pathway because no proton translocation occurs; protons required for the reduction of oxygen are believed to be taken from the matrix (N-phase) and so it follows that the oxidation of ubiquinol must release protons to the matrix rather than to the P-phase. The molecular nature of the alternative oxidase has long eluded definition although at the time of writing there is evidence that one or more polypeptides of around 35 kDa are components and the genes are being sequenced. A puzzling feature of the alternative oxidase is that no redox groups have been firmly identified even in partially purified preparations.

What is the physiological function of the alternative oxidase? In the case of mitochondria from Arum (often known as Lords and Ladies in the UK) it is to generate heat which will volatilize insect attractants to aid pollination. On

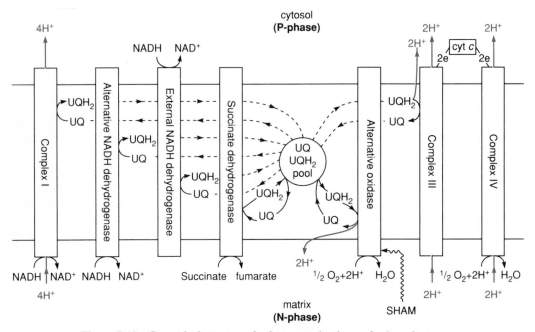

Figure 5.15 General features of the organization of the electron-transport system of plant mitochondria.
The role of a ubiquinone pool in connecting the various pathways of electron transport is evident. Adapted from Moore and Siedow, (1991).

the other hand, in the American skunk cabbage the heat production is directed towards permitting growth in sub-zero temperatures. This thermogenic mechanism offers a striking contrast with that evolved by mammalian brown fat mitochondria (Section 4.7.1), in which proton translocation is normal, but a dissipative proton re-entry pathway exists. A third rationale for the alternative oxidase pathway is that it provides a mechanism for oxidative metabolism in the absence of ATP synthesis. Finally, it is important to appreciate that the extent of activity of the alternative oxidase pathway varies between different plants. In potato, for example, it is present in only low levels, in contrast to the two plants discussed above, where it is the dominant pathway. The mechanism whereby electrons are distributed between two electron transport pathways to oxygen is not fully understood. In principle, it could depend on collisional interactions between ubiquinol and the redox proteins but detailed analysis has suggested that one or both of the pathways may be regulated in response to the ubiquinol/ubiquinone ratio.

Whereas the mammalian mitochondrial respiratory chain is unable to oxidize cytoplasmic NADH directly (Section 8.4.8), plant mitochondria usually possess a distinct inner membrane rotenone-insensitive NADH dehydrogenase in which the active site is exposed to the cytoplasm. Transfer of electrons from this enzyme to ubiquinone is not associated with the translocation of charge across the membrane. (There may be a third enzyme that oxidizes matrix NADH but which neither translocates protons nor is inhibited by rotenone (Fig. 5.15).)

An interesting example of the disruption of mitochondrial energy transduction, with resulting losses to agriculture of the order of $1 billion in the USA in past years, is provided by the effect of a toxin produced by the fungus *Bipolaris* (formerly *Helminthosporium*) *maydis*. This fungus causes leaf blight in male sterile corn plants with Texas male cytoplasm (Levings, 1990). The toxin, which is a polyketide, causes the inner mitochondrial membrane to become leaky to ions and small molecules. There may also be an inhibitory effect on the rotenone-sensitive NADH dehydrogenase (complex I). The toxin is ineffective against mitochondria from fertile and normal plants. The basis for this difference has been shown to reside in the presence in the susceptible mitochondria of a hydrophobic 13-kDa membrane protein which probably forms three membrane-spanning α-helices. Introducing the gene for this polypeptide into *E. coli* or yeast makes the cytoplasmic membrane of the former and the inner mitochondrial membrane of the latter susceptible to the toxin.

Neurospora crassa is a fungus which is widely used for biochemical investigations including mitochondrial biogenesis. Although the standard mitochondrial electron-transport system is present, there is also an alternative oxidase pathway. In common with the plant system this is inhibited by SHAM and the similarity between the two is further emphasized by the finding that monoclonal antibodies against the plant alternative oxidase cross-react with mitochondrial inner membranes from *N. crassa*. Trypanosome mitochondria may also possess this type of oxidase because cross-reaction with antibodies and SHAM sensitivity have been observed.

The SHAM-sensitive oxidase is not found in all types of fungal mitochondria. It has not, for example, been identified in yeast. There are, however, certain activities of yeast mitochondria which are not found in mammalian mitochondria. These include an NADH dehydrogenase for which the active site faces the cytoplasm, an L-lactate–cytochrome c oxidoreductase and a cytochrome c peroxidase. The latter two are located in the intermembrane space. In the oxidoreductase, electron transfer occurs from lactate via FMN and from b-type haem to cytochrome c which can transfer electrons to either cytochrome aa_3 or the peroxidase. The latter contains b-type haem, but a second redox active group is a specific tryptophan side chain (Prince and George, 1990). This is one of the relatively rare instances of an amino acid side chain undergoing an oxidation–reduction reaction as part of an electron-transport process; another example is given in Section 6.4.2.

5.13 BACTERIAL RESPIRATORY CHAINS

Reviews   Anraku 1988, Anthony 1988, Ferguson 1987a

Oxidative phosphorylation is important in many genera of bacteria (but not of course in organisms that exist by fermentation alone). As explained in Chapter 1, the study of bacterial oxidative phosphorylation required the development of cell-free vesicular systems. Their availability, together with the general developments in methodologies for characterizing membrane-bound redox proteins, means that a great deal is now known about electron transport in bacteria. The topic is especially large, not only because there are many different sources and acceptors of electrons can be used by various organisms, but also the components involved in electron transfer from a donor to a given acceptor can differ markedly between organisms, or even within the same organism depending on the growth conditions. Nevertheless, some general principles can be discerned, and attention has focused more strongly on bacterial electron transport in recent years because cloning, sequencing and expressing of genes is generally easier than for mitochondrial systems.

We shall restrict our discussion to a limited number of bacterial electron-transfer chains which have either been intensively investigated or which provide novel mechanistic insights.

5.13.1 *Paracoccus denitrificans*

This soil organism allows us to start on familiar ground since many features of its electron-transport system are similar, whatever its growth mode, to their mitochondrial counterparts (Fig. 5.16). There is, however, an intriguing difference: complexes II, III and IV all contain fewer polypeptide chains than their mitochondrial counterparts, facilitating their structure–function analysis. Unfortunately the *P. denitrificans* counterpart to complex I has not yet

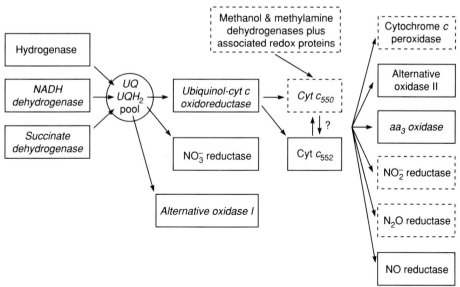

Figure 5.16   Organization of electron transport components in *P. denitrificans*.
Only the components in italics are thought to be constitutive. The other components are induced by appropriate growth conditions and are unlikely to be all present at once. NADH dehydrogenase, succinate dehydrogenase, ubiquinol cytochrome *c* oxidoreductase and *aa*$_3$ oxidase correspond to mitochondrial complexes I to IV. Continuous boxes indicate integral membrane components; dashed lines represent periplasmic components. Further details of methanol and methylamine oxidation are given in Fig. 5.17.

been isolated, although antibody cross-reactivity and some gene sequence data (Xu *et al.*, 1991) indicate that it is a closely related complex.

The cyt c_{550} (Fig. 5.16) is closely related to mitochondrial cyt *c* in terms of both structure and redox potential. It would thus be expected to shuttle between the proteins corresponding to the bc_1 and aa_3 complexes at the outer (P) surface of the cytoplasmic membrane. However, a membrane-bound cyt c_{552} may instead carry out this role: thus deletion of the cyt c_{550} gene does not stop electron transfer to cyt aa_3. We return to this unexpected complication later.

Cyt aa_3 does not provide the only route to oxygen (Fig. 5.16), since there is an alternative which bypasses both bc_1 and aa_3 as well as a probable third oxidase that can accept electrons from *c*-type cytochromes. The molecular nature of these alternative oxidases and the factors that regulate their synthesis and activity are not understood. The pathway that branches at UQ has a lower $H^+/2e^-$ stoichiometry than the bc_1/aa_3 pathway. Such branching and the lack of understanding of the reasons for it is typical of bacterial electron-transport chains. Figure 5.16 also shows a hydrogenase (different from the NAD^+-dependent *A. eutrophus* hydrogenase that is related to NADH dehydrogenase – Section 5.6) that can pass electrons directly to the UQ pool. Electrons can also be fed to UQ from succinate dehydrogenase as well as from a FAD-linked fatty acid oxidation step (not shown in Fig. 5.16) (c.f. Fig. 5.1).

P. denitrificans can use final electron acceptors other than oxygen. Amongst these is H_2O_2, a molecule commonly found in the soil environment of the bacteria. Reduction of H_2O_2 is catalysed by a periplasmic cytochrome c peroxidase (Fig. 5.16) which is a dihaem c-type cytochrome and thus differs from the enzyme with the same function that is in the intermembrane space in yeast mitochondria (Section 5.12).

(a) Oxidation of compounds with one carbon atom

P. denitrificans is one of a restricted group of bacteria that can grow on methanol or methylamine as sole carbon source. The dehydrogenases for these two compounds (Fig. 5.17) are found in the periplasm (P-phase). That for methanol contains pyrroloquinoline quinone (PQQ) (Duine, 1991) as a cofactor whilst methylamine dehydrogenase (for which a three-dimensional structure has been determined) contains a novel type of redox centre, a tryptophyl–tryptophan involving a covalent bond between two tryptophan side chains (McIntire et al., 1991). Electrons pass from the redox centre of methylamine dehydrogenase to a one-electron carrier copper protein, amicyanin, which forms a sufficiently tight complex, mainly through apolar interactions (Matthews et al., 1991), with the dehydrogenase to allow co-crystallization. From amicyanin the electrons pass to c-type cytochromes, probably including cyt c_{550}, and then to oxygen via cytochrome aa_3 oxidase. Thus Δp is established in this short electron-transfer chain by the inward movement of electrons and outward pumping of protons through cyt aa_3

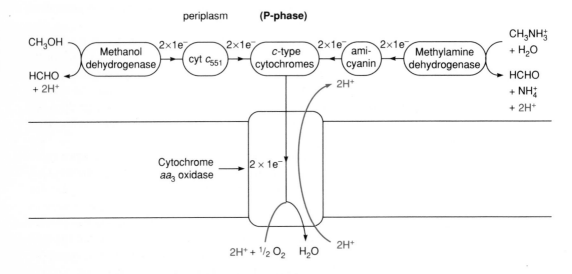

Figure 5.17 Schematic representation of periplasmic oxidation of methanol or methylamine in P. denitrificans.

together with the release and uptake of protons at the two sides of the membrane associated with methylamine oxidation and oxygen reduction (Fig. 5.17).

In the case of methanol oxidation the energetic considerations are similar, electrons being transferred from reduced PQQ to c-type cytochromes (Fig. 5.17). Although the $E_{m,7}$ of the methanol/formaldehyde couple is similar to that for succinate/fumarate, the H^+/O stoichiometry for the oxidation of succinate (which involves both the bc_1 complex and cytochrome aa_3 oxidase) is greater than for methanol oxidation. Why a longer electron-transport chain is not used for the oxidation of methanol or methylamine is not understood. The formaldehyde produced from oxidation of methanol or methylamine is oxidized by cytoplasmic (N-phase) enzymes to CO_2 with concomitant generation of NADH. In *P. denitrificans* the CO_2 thus produced is refixed into cell material. In other organisms that grow on methanol some of the formaldehyde can be directly incorporated into cell material (Gottschalk, 1986).

(b) Denitrification

Review   Ferguson 1987b

The sequential reduction of NO_3^-, NO_2^-, NO and N_2O is catalysed by anaerobically grown *P. denitrificans* in the process known as denitrification (hence the name of the organism). The four reductases required to carry out this process receive electrons from the underlying electron-transport system used in aerobic respiration (Fig. 5.16 and 5.18).

The membrane-bound NO_3^- reductase is thought to receive electrons from UQH_2 on the periplasmic (P) side of the membrane. $2H^+/UQH_2$ are released to the periplasm and two electrons pass inwards across the cytoplasmic membrane via two b-type haems to the site of NO_3^- reduction on the cytoplasmic (N) surface (Fig. 5.18). The $E_{m,7}$ for the NO_3^-/NO_2^- couple is $+420$ mV. The inward movement of the electrons is equivalent to the transfer from cytoplasm to periplasm of two positive charges ($2q^+$) per 2 e^-. Since UQ was originally reduced at the N face of the membrane, the outward transfer of UQH_2 and the return to the N-phase of two electrons can be considered as an example of a classic Mitchell 'loop' (Fig. 4.9).

NO_2^- reductase is a soluble enzyme in the periplasm, contains both c- and d_1-type haems and can receive electrons from cyt bc_1 via cyt c_{550} on the periplasmic face of the membrane. The transfer of two electrons through bc_1 to the periplasmic cyt c_{550} coupled to the release to the periplasm of $4H^+/2e^-$ is, as in the case of the mitochondrial bc_1 complex, equivalent to $2q^+/2e^-$. (Note that these nitrate and nitrite reductases, and the *E. coli* enzymes discussed below (Section 5.13.12), are distinct from the widespread enzymes with the same names that are responsible for the assimilation of nitrogen in bacteria and plants and which are beyond the scope of this book.)

NO reductase is an integral membrane protein containing both b- and c-type haems. The energetics of nitric oxide reduction are not fully characterized but NO is probably reduced at the periplasmic surface of the membrane by electrons again supplied from UQH_2 via cyt bc_1 and cyt c_{550}, and so as in the case of NO_2^- reductase the net outward charge transfer is $2q^+/2e^-$.

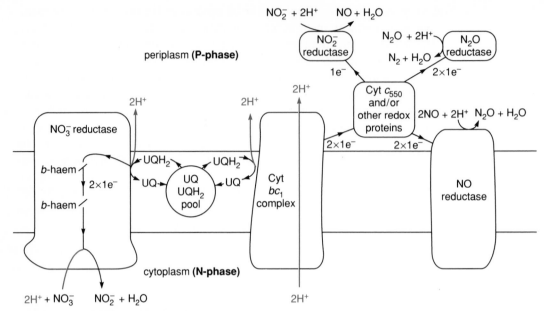

Figure 5.18 Electron-transport pathways associated with denitrification in *P. denitrificans*.
The flow of electrons to nitrate from ubiquinol occurs through the reductase which is proposed to have two *b*-type haems distributed across the membrane and its molybdenum centre, where nitrate binds, in the cytoplasmic domain of the enzyme. As explained in the text, this loop mechanism is associated with the same net positive charge translocation across the membrane as electron flow to the other three nitrogenous acceptors. The charge and proton translocation stoichiometry catalysed by the ubiquinol cytochrome *c* oxidoreductase is explained in Section 5.8 and Fig. 5.12.

Finally N_2O reductase, a copper-containing enzyme, is, like NO_2^- reductase, a soluble periplasmic enzyme linked to bc_1 via c_{550} and associated with a $2q^+/2e^-$ outward charge transfer (Fig. 5.18). It is notable that the $E_{m,7}$ of the N_2O/N_2 couple is even more positive then $\frac{1}{2}O_2/H_2O$, namely $+1100$ mV, although the concentration of N_2O *in vivo* may be so low that the actual E_h for the couple may be comparable to the $+800$ mV for the latter.

Nitrite reductase, N_2O reductase and NO reductase thus all appear to serve as P-face electron sinks at the level of cyt *c* (Fig. 5.18), and as such all play integral roles in their respective pathways. Each is associated with $2q^+/2e^-$ stoichiometries for the spans from UQH_2, as indeed is NO_3^- reductase with its different pathway. Thus despite the redox span from the UQH_2/UQ couple to NO_3^-/NO_2^- being far smaller than to N_2O/N_2 the charge transfer is the same (Fig. 5.18). It should be noted also that the onward flow of electrons from cyt c_{550} to oxygen via cyt aa_3 rather than to these reductases leads to a higher $q^+/2e^-$ ratio.

There has been some confusion as to whether a soluble periplasmic enzyme such as N_2O reductase can 'participate' in the generation of Δp. It should be clear that although the activity of such enzymes *per se* does not contribute directly to Δp, their roles in transferring electrons to acceptors is necessary for the proton-translocating NADH dehydrogenase and bc_1 complex to function. On the other hand, comparison of Fig. 5.17 with Fig. 5.18 shows that transfer of electrons from the periplasmic methanol dehydrogenase to N_2O reductase would not generate a Δp despite the large redox span. This illustrates the importance of considering not only redox spans but also the topology of electron flow in energy-transducing membranes (Ferguson, 1987a). One complication in the scheme of Fig. 5.18 is that the role of cyt c_{550} in denitrification has recently been thrown in doubt by the finding that a mutant of *P. denitrificans* lacking cyt c_{550} is still able to denitrify. A tentative explanation is that a second electron-transport protein can substitute for c_{550}.

(c) *Relationship to other bacterial systems*

Many of the features of the electron-transport system of *P. denitrificans* are of general importance. First we see that the UQ/UQH_2 pool not only acts as a collector of electrons from several sources, as in mitochondria, but also donates electrons to several alternative pathways, and thus serves as a 'crossroads' for electron transfer, compatible with its mobility within the bilayer. It is not known how electrons distribute themselves between the different pathways; the simplest mechanism would be competition, which may also occur in plant mitochondria (Section 5.12). In some cases this explanation appears insufficient; thus nitrate reduction by *P. denitrificans* is blocked by the presence of oxygen. The locus of control may be a nitrate transport system.

The *c*-type cytochromes, which in bacteria are far more varied and play a greater range of roles than in mitochondria, are also a common feature of bacterial electron transport (Pettigrew and Moore, 1987). Such cytochromes are often water soluble and almost invariably found in the periplasm or at least with a haem group exposed to the periplasm (e.g. cyt c_1). The presence in the periplasm of a large number of electron-transfer proteins is also a general feature of redox reactions in Gram-negative bacteria. In Gram-positive bacteria, which do not have a periplasm, the *c*-type cytochromes appear to be more tightly associated with the cytoplasmic membrane and the range of metabolic activities associated with periplasmic dehydrogenases and reductases is much more restricted than in Gram-negative organisms.

Some of the electron-transport chain components found in *P. denitrificans*, especially bc_1 (Trumpower, 1990b) and aa_3, are also relatively widely distributed, although aa_3 can be absent when bc_1 is present. In some denitrifying bacteria nitrite reductase does not contain c and d_1 haems but rather is a copper protein (a high-resolution structure shows a trimeric protein with type I copper as in plastocyanin (Section 6.4) and type II copper as the active site (Godden *et al.*, 1991)) and we shall see in the next section that the most familiar bacterium in experimental biochemistry, *E. coli*, has a distinct set of components for catalysing electron flow to oxygen.

5.13.2 *Escherichia coli*

Reviews   Anraku and Gennis 1987, Poole and Ingledew 1987, Lin and Kuritzkes 1987

When *E. coli* is grown aerobically the electron-transport components are quite distinct from those found in either mammalian mitochondria or *P. denitrificans*. In particular, there are no detectable *c*-type cytochromes and electron transport from UQH_2 to oxygen is insensitive to antimycin and myxothiazol, inhibitors of the bc_1 complex; the latter is absent from *E. coli*. Two oxidase enzymes, known as cyt *bo* and cyt *bd*, can directly oxidize ubiquinol (Fig. 5.19). Cyt *bo* comprises four polypeptides and contains Cu (probably 2 atoms) and two *b*-type haems, one of which reacts with O_2 and is designated as the *o* centre. This cyt *bo* complex is, in common with cytochrome aa_3, a proton pump with stoichiometry $2H^+/2e^-$. Unexpectedly, three of the subunits in cyt *bo* have some striking sequence similarities with the cyt aa_3 complexes from mitochondria (complex IV) and other bacteria. A Cu_A centre is not present but there is evidence for a bimetallic haem–copper centre possibly similar to the haem a_3–copper centre in cytochrome aa_3 oxidase.

The cyt *bd* complex comprises two polypeptide chains and has two *b*-type haems in addition to the distinctive porphyrin ring of the *d*-type haem (different from the d_1 haem in *P. denitrificans* nitrite reductase (Section 5.13.1a)) which is the site of oxygen reduction. There is no evidence that cyt *bd* is a proton pump and thus the stoichiometry of charge translocation is $2q^+/2e^-$, due purely to the inward movement of electrons from the site of ubiquinol oxidation, in other words a loop mechanism (cf. Fig. 4.9). Cyt *d* has a much higher affinity for oxygen than cyt *bo* and is synthesized under conditions of low oxygen concentrations. The lower stoichiometry of proton translocation achieved at low oxygen concentrations may be the price that has to be paid to attain a high catalytic rate of oxidase activity with no thermodynamic back-pressure from the protonmotive force on the individual reaction steps of oxygen reduction (Puustinen *et al.*, 1991). A similar cytochrome *bd* oxidase with high affinity for oxygen terminates a ubiquinol oxidase system in *Azotobacter vinelandii* and *Klebsiella pneumoniae*. In these organisms a role

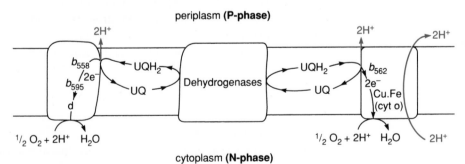

Figure 5.19   The *E. coli* aerobic electron-transfer chain from ubiquinol to oxygen.
Adapted from Anraku and Gennis (1987)

of this oxidase is to maintain low oxygen concentrations in order to protect the oxygen-sensitive nitrogenase.

Clearly *E. coli* has a truncated electron-transport chain, in comparison with mitochondria and *P. denitrificans*, with lower $q^+/2e^-$ and $H^+/2e^-$ ratios. If the H^+/ATP ratio for the ATP synthase is the same (i.e. 3) as for other organisms then it follows that the maximum P/O ratio for $UQH_2 \rightarrow \frac{1}{2}O_2$ is also lower (0.67 for electron flow to cyt *d* and 1.33 for cyt *bo* instead of the theoretical value of 2 for sub mitochondrial particles and 1.5 for mitochondria). Obviously the redox span for $UQH_2 \rightarrow \frac{1}{2}O_2$ is independent of the pathway. The electron-transport system of *E. coli* illustrates that an organism may not always be seeking to maximize the stoichiometry of ATP production. Natural habitats of *E. coli* may be rich in potential substrates and the need to maximize ATP yield may not apply.

The NADH dehydrogenase of *E. coli* is not well characterized (in common with the enzyme from other bacteria). Although one form of the enzyme can translocate protons, it is not as closely related to the mitochondrial complex I as its counterpart in *P. denitrificans*.

E. coli is not restricted to aerobic growth. Under anaerobic conditions part of the citric acid cycle, 2-oxoglutarate dehydrogenase, ceases to function (Spiro and Guest, 1991), in contrast to *P. denitrificans* and many other non-enteric bacteria. Pyruvate can be converted to formate or fumarate and consequently under anaerobic conditions these can respectively act as an electron donor and acceptor to the electron-transport system that contains menaquinone rather than ubiquinone under these conditions. Oxidation of formate

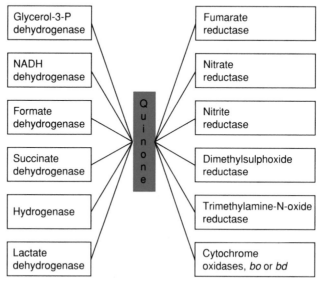

Figure 5.20   An overview of *E. coli* aerobic and anaerobic respiratory systems.
The components present depend on the growth conditions. Under anaerobic conditions menaquinone replaces ubiquinone as the main quinone.

to CO_2 and concomitant reduction of fumarate to succinate can generate Δp. Other anaerobic acceptors and donors can be used (Fig 5.20). The expression of many of the respective enzymes is dependent on the Fnr protein which under anaerobic conditions acts as a transcriptional activator for many of the genes associated with anaerobic metabolism. NO_3^- is reduced to NO_2^- by a reductase that is very similar to that described above for *P. denitrificans*. However, NO_2^- is reduced to NH_4^+, rather than to NO, by a periplasmic nitrite reductase that contains six *c*-type haem groups. Dimethylsulphoxide and trimethylamine-*N*-oxide (both occur in natural environments, the latter especially in fish) can also serve as terminal electron acceptors via one or more reductases for these molecules (Fig. 5.20).

5.13.3 *Nitrobacter* and *Thiobacillus ferroxidans*

If an organism grows on a substrate with a relatively positive redox potential it can be faced with the problem of how to generate NADH or NADPH for biosynthetic reactions. The example of *Nitrobacter* is taken here to illustrate this aspect of electron transport (Ferguson, 1987a).

Nitrobacter grows by oxidizing nitrite to nitrate ($E_{m,7} + 420$ mV) using a nitrite oxidoreductase, transferring electrons via a *c*-type cytochrome to a cyt aa_3 oxidase and reducing oxygen to water ($E_{m,7} + 820$ mV), Fig. 5.21. It is not immediately apparent, therefore, how the organism can reduce NAD^+ to NADH ($E_{m,7} - 320$ mV) for biosynthetic reactions. The solution comes from reversed electron transfer, which was introduced in a mitochondrial context in Section 4.11.

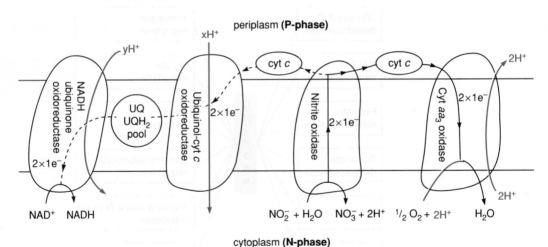

Figure 5.21 Protonmotive force generation and reversed electron transport in *Nitrobacter*.

As explained in the text it is proposed that $\Delta\psi$ drives electrons energetically uphill from nitrite to the cytochrome *c* and that cytochrome aa_3 acts as a proton pump.(------) indicates reversed electron flow. Not all details of this scheme have been fully substantiated.

The short electron-transfer chain described above generates a Δp which will, as in other bacterial genera, drive ATP synthesis and transport processes. *Nitrobacter* probably possesses a bc_1 complex (or an equivalent) and an NADH dehydrogenase complex which, as in other electron-transfer chains, are reversible. The Δp generated by NO_2^-/O_2 is used to reverse both these complexes and proton re-entry drives a minority of the electrons originating from NO_2^- back through the complexes from c-type cytochrome to NADH (Fig. 5.21).

The mechanism of Δp generation has not been fully elucidated. Figure 5.21 presents a plausible, but not fully proven, scheme in which NO_2^- oxidation occurs on the N-face of the membrane, transferring electrons to cyt c on the P face. This may seem strange, since this charge movement *collapses* rather than generates Δp, but there is a thermodynamic reason. The $E_{m,7}$ for the cytochrome is $+270 \, mV$, or $150 \, mV$ more electronegative than for NO_2^-/NO_3^-. Δp, or rather the $170 \, mV$ $\Delta \psi$ which is typical for bacterial cytoplasmic membranes, is needed to drive the reduction of the cytochrome. A related utilization of $\Delta \psi$ has been discussed in the context of the mitochondrial bc_1 complex (Section 5.8). From cyt c electrons pass to oxygen via the proton-pumping cytochrome aa_3 oxidase. Note that the electrons return to the N-face and that the net outward movement of positive charge is due to the proton pumping of the oxidase.

Support for the above role of the membrane potential comes from studies with inverted membrane vesicles from *Nitrobacter*. Electron transfer from NO_2^- to O_2 is slowed, rather than accelerated as in mitochondrial oxidations (Section 4.6), by conditions which decrease $\Delta \psi$ (e.g. presence of protonophores or 'state 4/state 3' transition (Section 4.6)). Such conditions result in a decreased reduction of cyt c and hence a decline in respiration. The energetics of *Nitrobacter* illustrate the beautiful economy of the chemiosmotic mechanism. Δp drives the initial step of substrate oxidation and reversed electron transport as well as more conventional processes such as ATP synthesis and substrate transport.

Nitrobacter is by no means the only example of an organism in which reversed electron transport is important. Another instance is *Thiobacillus ferroxidans* which oxidizes Fe^{2+} to Fe^{3+} ($E_{m,7} = +780 \, mV$) (Ingledew, 1982; Yamanaka *et al.*, 1991). As with the oxidation of nitrite to nitrate, this reaction cannot directly reduce NAD^+ and thus a small proportion of the electrons derived from Fe^{2+} are transferred 'uphill' to NAD^+ whilst the remainder flow to oxygen with concomitant generation of Δp. The oxidation of Fe^{2+} occurs in the periplasm and electrons may be transferred to an oxidase via a copper protein known as rusticyanin (Fig. 5.22).

5.13.4 The bioenergetics of methane synthesis by bacteria

Reviews Blaut *et al.* 1990, Thauer, 1990

Methanogenic bacteria are archaebacteria which obtain energy from several

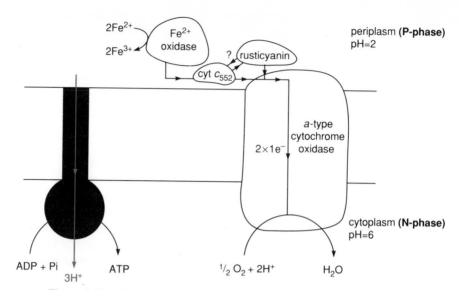

Figure 5.22   Electron transfer by *Thiobacillus ferroxidans*.
The large pH gradient and the small redox span between Fe^{2+} and O_2 means that the cytochrome oxidase cannot additionally pump protons across the membrane. The inward movement of electrons through the oxidase will contribute to generation of a $\Delta\psi$, negative inside. This charge movement will also balance the inward movement of protons through the ATP synthase. (Note that if the site of oxygen reduction were to be on the periplasmic side the formal redox span would increase because of the pH dependence of the O_2/H_2O couple, but there would be no gain in the energy available because protons required for the reduction would have to be brought from the cytoplasm uphill against their concentration gradient. Periplasmic reduction of O_2 together with use of protons derived from the periplasm would not contribute to the generation of a Δp.) This scheme indicates that less than 1 mol ATP is synthesized per 2e flowing from Fe^{2+} to O_2.

types of reaction in which methane is an end-product. It has only recently been established that this methane formation is associated with electron-transport-driven H^+ or Na^+ translocation and that resultant ATP synthesis is by a chemiosmotic mechanism.

Two methanogenic organisms, *Methanosarcina barkeri* and a species known only as *strain Gö*1, which can both gain energy for growth from the reduction of either CH_3OH or CO_2 by H_2, have provided important clues about the bioenergetics of methanogenesis:

$$CH_3OH + H_2 \rightarrow CH_4 + H_2O$$

or:

$$CO_2 + 4H_2 \rightarrow CH_4 + 2H_2O$$

The organisms can also grow on methanol alone, but discussion of this will be reserved until the fundamental electron pathways have been described for reduction of CH_3OH and CO_2.

(a) Reduction of CH_3OH to CH_4 by H_2

Intact cells of *M. barkeri* can synthesize ATP as well as CH_4 in the presence of CH_3OH and H_2. ATP synthesis is chemiosmotic (rather than a result of substrate level phosphorylation by a soluble enzyme system) by the following criteria:

- Protons are extruded.
- DCCD (dicyclohexycarbodiimide), presumed to be a specific inhibitor of an ATP synthase as in other organisms, inhibits ATP production, increases Δp and slows the rate of methane formation.
- Protonophores dissipate Δp but increase the rate of methane formation.

These observations parallel what would be observed in the analogous mitochondrial proton circuit (Chapter 4). An involvement of Na^+ can be eliminated since this ion is not needed for methanogenesis from CH_3OH plus H_2 although Na^+ is required for growth of methanogens and some reactions of methanogenesis (see later).

Further understanding required an organism, known as *strain Göl*, from which functional inside out membrane vesicles could be isolated. Addition of H_2 and CH_3SCoM (which is formed from CH_3OH and coenzyme M ($HSCH_2CH_2SO_3^-$, CoM) in a corrinoid-dependent process) to crude Göl vesicles drove ATP synthesis. The methyl group of CH_3SCoM is converted to methane through reaction with a second thiol-containing compound known as HTP-SH to give a heterodisulphide:

$$CH_3SCoM + HTP-SH \rightarrow CH_4 + CoM-S-S-HTP$$

The enzyme catalysing this reaction requires as a prosthetic group a Ni-containing porphinoid known as F_{430}. However, it is not the above reaction (step 7 in Fig. 5.23) that directly drives ATP synthesis because it is the reduction of $CoM-S-S-HTP$ back to the two separate thiol species (step 8) that is catalysed by a membrane-bound enzyme coupled to proton translocation. An electron source for this reduction is reduced coenzyme F_{420} (to be distinguished from F_{430}) from which electrons are transferred by unknown redox centres that probably constitute a proton-translocating electron transport chain (Fig. 5.23). F_{420} is a $5'$-deazaflavin with a $E_{m,7}$ of -370 mV which is a structural and functional hybrid between nicotinamide and flavin coenzymes and occurs as a diffusible species in the cytoplasm of methanogenic bacteria. In line with a chemiosmotic mechanism the rate of oxidation of reduced F_{420} is accelerated by onset of ATP synthesis or addition of protonophores. F_{420} is re-reduced by electrons from H_2 in a reaction catalysed by a specific hydrogenase.

(b) Reduction of CO_2 to CH_4 by H_2

Growth of *M. barkeri* is also supported by the reduction of CO_2 by H_2. CO_2 is first taken up by covalent attachment to methanofuran (Fig. 5.27). After a first reduction step to a formylated derivative, using electrons derived from

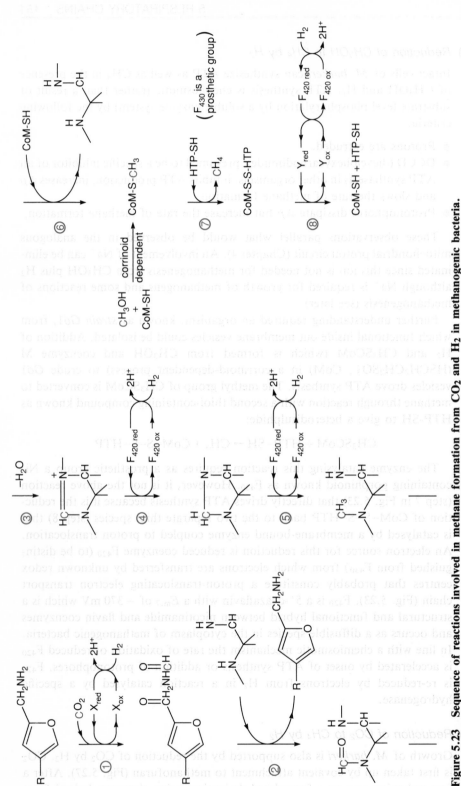

Figure 5.23 Sequence of reactions involved in methane formation from CO₂ and H₂ in methanogenic bacteria.

As explained in the text, reaction step 1 is endergonic whereas steps 5, 6 and 8 are exergonic. X is the unknown immediate donor of electrons in step 1; there is an F₄₂₀-independent hydrogenase which may donate to X. An F₄₂₀-dependent hydrogenase would provide the reduced F₄₂₀ for steps 4, 5 and 8. Electrons destined for step 8 are shown as passing through an uncharacterized series of electron carriers (Y) from a dehydrogenase for reduced F₄₂₀. Electrons from methanol enter the sequence as shown, and then in disproportionation of methanol (see Section 5.13.4c) the electrons flow in a 3:1 proportion to give either methane, with concomitant Δp generation, or to yield CO₂ by the reverse of reactions 6 to 1. With H₂ + CH₃OH (see Section 5.13.4a) all the CH₃SCoM is converted to CH₄. Study of methanogenesis is continuing and some details of this scheme may need revision in due course. Note that the text and figure do not deal with the main *in vivo* route of methanogenesis, from acetate. See Blaut *et al.* (1990), Thauer,

H_2, this formyl group is transferred to a pterin compound. After two further reductions in which electrons from H_2 are transferred via F_{420}, a methyl group is formed which can be transferred to coenzyme M (see above and Fig. 5.23). The CH_3SCoM then reacts as described above (Section 5.13.4a) to generate CH_4.

(c) Growth by disproportionation of CH₃OH

Growth of *M. barkeri* and *Göl* can grow by disproportionation of methanol in the absence of H_2:

$$4CH_3OH \rightarrow 3CH_4 + CO_2 + 2H_2O$$

The stoichiometry of this reaction shows that one molecule of methanol is used to provide the reductant required for methane formation from the other three molecules of methanol. ^2H-labelling shows that three hydrogens in each of the three methane molecules are derived from the methyl groups in CH_3OH, probably via CH_3SCoM. The electrons released in the oxidation of the fourth methane are, of course, those needed for the reduction of three molecules of CH_3SCoM to CH_4. It is very likely that CH_3SCoM is converted to CO_2 by the reverse of the reactions shown in Fig. 5.23 , except that the formation of hydrogen, rather than reduced F_{420}, must be avoided since the electrons must instead drive the reductive reaction of methane formation (Step 8 in Fig. 5.23).

(d) The energetics of methanogenesis

Although Fig. 5.23 summarizes the likely electron-transfer steps involved in the reduction of CO_2 to CH_4 it gives no information about the bioenergetics of the process except that we have already established that reduction of $CoM–S–S–HTP$ is coupled to proton translocation (Section 5.13.4a) and this would allow the cells to grow on CH_3OH alone. Are there any energy-conserving steps in the more complex reduction of CO_2? Analysis of the thermodynamics of the individual steps suggests that the reduction of the methylene derivatives of the pterin and/or the following methyl transfer reaction (step 5 and 6 in Fig 5.23) are exergonic. Reduction of added formaldehyde (which attaches spontaneously to the pterin so as to enter the sequence after step 4) resulted in the generation of an electrochemical gradient although Na^+, rather than H^+, was translocated out of the cells. The exact role of the Na^+ gradient thought to be generated by step 5 and/or 6 (Fig. 5.23) is unclear, but it should be noted that methanogenic bacteria generally require Na^+ for growth.

At physiological concentrations of H_2, which are very considerably below the standard state value of 1 atmosphere, the first reaction (step 1 in Fig. 5.23) is significantly endergonic. It is therefore very probable that this step is driven by the inward movement of Na^+ (or H^+) down the electrochemical gradient set up by reactions 5 and/or 6 and 8.

Whereas $\Delta G^{0\prime}$ for the reduction of CO_2 by H_2 is -131 kJ mol^{-1} per mol CH_4 produced, the actual $\Delta G'$ is likely to be closer to -30 kJ mol^{-1}. Since

ΔG_p, the free energy for ATP synthesis, is likely to be about $+50 \text{ kJ mol}^{-1}$, this means that a theoretical maximum of about 0.6 ATP can be generated per mol CH_4 formed from CO_2. This is a further example of the chemiosmotic mechanism allowing non-integral stoichiometries (see Chapter 4).

5.13.5 *Propionigenium modestum*

Review Dimroth 1991

We next turn to an example of bacterial energy transduction that does not involve electron transport but which, in common with electron-transport-dependent energy transduction, involves the co-operation of two ion pumps, and so is appropriately discussed in this chapter. *P. modestum* is an anaerobic bacterium that ferments succinate to propionate by a short reaction sequence:

succinate → succinyl CoA → methylmalonyl CoA →

propionyl CoA → propionate.

The decarboxylation of methylmalonyl CoA has a K' of about 10^5 ($\Delta G^{0\prime} = 27 \text{ kJ mol}^{-1}$), close to that for the overall fermentation. The actual ΔG is likely to be fairly close to this value. The decarboxylase is a membrane-bound, biotin-dependent enzyme that pumps two Na^+ out of the cell for each CO_2 released. The Na^+ electrochemical gradient thereby set up could be up to 12–15 kJ mol^{-1} at equilibrium (since two ions are pumped) and is known to drive ATP synthesis through a Na^+-translocating ATP synthase that is discussed further in Chapter 7. Since a typical ΔG_p in a bacterial cell might be 45–50 kJ mol^{-1}, it is expected that three, or more likely four, Na^+ ions might be required per ATP synthesized for energetic reasons. The energetics of this organism reinforce the significance of non-integral coupling stoichiometries in bacterial energetics; this organism could not by definition exist if one ATP had to be formed by a soluble enzyme system for each molecule of succinate fermented! This is also not the only example of bacterial energy conservation being linked to a decarboxylation reaction. A further example is given in Section 8.5.5.

Spinach chloroplasts are subjected to an acid bath (Figure 4.20) in order to generate ATP in the dark, while the "Z" scheme of non-cyclic electron transfer generates O_2 and transfers electrons to a negative E_m component

6 PHOTOSYNTHETIC GENERATORS OF PROTONMOTIVE FORCE

6.1 INTRODUCTION

A photosynthetic organism captures light energy in order to drive the otherwise endergonic synthesis of molecules needed for the growth and maintenance of the organism. A central feature of photosynthesis is the conversion of light energy into redox energy, meaning that photon capture causes a component to change its redox potential from being relatively electropositive to being highly electronegative. The electrons released from this component are utilized to generate a Δp, flowing either through a cyclic pathway back to rereduce the original component or in a noncyclic pathway to reduce additional electron acceptors (ultimately $NADP^+$ in the case of thylakoid photosynthesis). In this latter case a continual electron supply to the photon-sensitive component is required (obtained from H_2O in the thylakoid example).

The production of ATP by photosynthetic energy-transducing membranes involves a proton circuit which is closely analogous to that already described for mitochondria and respiratory bacteria. Thus a Δp in the region of 200 mV across a proton-impermeable membrane is used to drive a proton-translocating ATPase in the direction of ATP synthesis. (In the case of photosynthetic bacteria Δp may also drive other endergonic processes (Fig. 1.9) – including reversed electron transport to generate NADH, see Section 6.3). The ATPase (or ATP synthase) is identical to the mitochondrial enzyme except in detail (Section 7.2). The distinction between the respiratory and photosynthetic systems is in the nature of the primary generator of Δp, yet even here a number of familiar components recur, including cytochromes, quinones and

Fe/S centres. Photosynthetic activity in halobacteria, (see Section 6.5) is distinctive: photon capture leads to a direct generation of Δp in the absence of electron transfer.

The two features that are unique to photosynthetic systems are the antennae, responsible for the trapping of photons, and the reaction centres, to which the light energy is directed. A component in the reaction centre becomes electronically excited as a result of the absorption of a photon. An electron can be released from this excited state at a potential which is up to 1 V more negative than the potential of donors to the reaction centre. Thus the electron lost from the reaction centre is replaced by an electron at much more positive potential so as to regenerate the ground state of the component in the reaction centre that underwent excitation. In this way light energy is directly transduced into redox potential energy. This sequence of photochemical events is often denoted by the shorthand

$$P \longrightarrow P^* \xrightarrow{-e^-} P^+ \xrightarrow{+e^-} P$$

where P indicates a pigment in the reaction centre, * an electronically excited state and P^+ the cation form of the ground state of the pigment. The separate components that accept the electron (e^-) from P^* and donate it to P^+ are described later.

In the case of the representative purple photosynthetic bacterium *Rhodobacter sphaeroides*, the electron released from the reaction centre feeds into a bulk pool of ubiquinol from which it passes via a proton-translocating

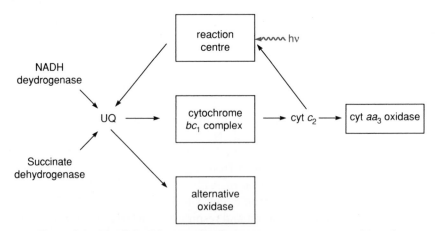

Figure 6.1 **The light-driven cyclic electron transport system and its relationship to respiratory electron transport in *R. sphaeroides*.**
This is a simplified version. Deletion of the gene for cytochrome c_2 does not prevent cyclic electron transport because an alternative *c*-type cytochrome can act as substitute. Electron transport in the closely related organism *Rhodobacter capsulatus* is similar except that cytochrome aa_3 is absent and replaced by another type of oxidase. Other aspects of electron transport in these two organisms, including anaerobic respiratory pathways, are given in Ferguson *et al.* (1987).

cyt bc_1 complex (Chapter 5) to a cyt c_2 (closely related to mitochondrial cyt c and *P. denitrificans* cyt c_{550} (Chapter 5)). Cyt c_2 in turn acts as the donor of electrons to the reaction centre, completing the cyclic electron flow (Fig. 6.1).

Whilst a cyclic electron-transfer pathway is also present in chloroplasts (see Section 6.3) they are distinguished by a non-cyclic pathway in which electrons are extracted from water, pass through a reaction centre, a proton-translocating-electron transfer chain (which again has similarities to the mitochondrial complex III), and a second reaction centre, and are ultimately donated to $NADP^+$, at a redox potential 1.1 V more negative than the $\frac{1}{2}O_2/H_2O$ couple (see Section 6.4). The chloroplast thus not only accomplishes an 'uphill' electron transfer but at the same time generates the Δp for ATP synthesis. The ATP and NADPH are used in the Calvin cycle, the dark reactions of photosynthesis in which CO_2 is fixed.

6.2 THE LIGHT REACTION OF BACTERIAL PHOTOSYNTHESIS

The heavily pigmented membranes of photosynthetic organisms act as antennae, absorbing light and funnelling the resultant energy to the reaction centres. In the case of *R. sphaeroides* the photochemically active pigment has an absorption maximum at 870 nm and thus is known as P_{870}. The equivalent energy of a 870-nm photon amounts to 1.42 eV (Section 3.5); thus the energy-transfer process is highly effective in the sense that 70% of the energy captured by the reaction centre is conserved in the resulting redox change of aproximately 1 V (Fig. 6.2).

6.2.1 Antennae

Review   Hunter *et al.* (1989)

The photochemical activity of reaction centres depends upon the delivery of light at a specific wavelength (870 nm for the commonly studied reaction centre of *R. sphaeroides*). Energy of this wavelength can be obtained by direct absorption of incident light with this wavelength and also by transfer, through mechanisms to be described below, from components of the reaction centre which absorb at shorter wavelengths. However, even in bright sunlight an individual pigment molecule will be hit by an incident photon only about once a second. Much higher rates of energy arrival at the reaction centre are required, to match the turnover capacity of the centre, which is $>100\,s^{-1}$. Some collecting process is evidently required.

Additionally, most of the incident photons will have the wrong wavelengths to be efficiently absorbed by the pigments of the reaction centre. Effective light absorption over a wide range of wavelengths shorter than 870 nm is achieved by an assembly of polypeptides with attached pigment molecules. These polypeptides, which bind bacteriochlorophyll and carotenoid pigments, are known

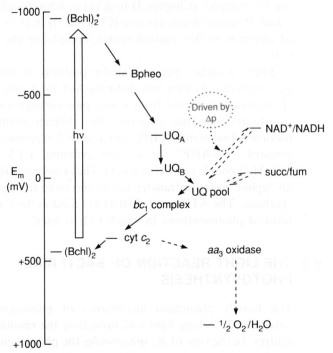

Figure 6.2 **Pathways of electron transfer in *R. sphaeroides* in relation to the redox potentials of the components.**

as light-harvesting or antenna complexes and surround the reaction centre. That such antenna are not strictly necessary for photosynthesis is established by the existence of bacterial mutants lacking them but which will nevertheless grow photosynthetically, albeit only in bright light.

Use of antennae to speed up the rate of photochemistry in the reaction centres is clearly more effective in biosynthetic terms than inserting very many copies of the reaction centre into the membrane to achieve a high rate of overall photochemistry. Thus over 99% of the bacteriochlorophyll molecules in a photosynthetic membrane are involved, together with carotenoid molecules, in absorbing light at shorter wavelengths than 870 nm (*R. sphaeroides*) and transferring it down an energy gradient to the lower energy absorption band at 870 nm. The transfer can occur by one of two mechanisms. The first, known as resonance energy transfer, is intermolecular, depending on an overlap between the fluorescence emission spectrum of a donor molecule and the excitation spectrum of an acceptor. Factors which affect the efficiency of such transfer include the relative orientation of donor and acceptor as well as the distance between the donor and acceptor (an inverse sixth power relationship). Energy transfer by this mechanism (which is not an emission followed by reabsorption of light) occurs over a mean distance of 2 nm in about 10^{-12} s.

At intermolecular separations of less than about 1.5 nm direct interactions between molecular orbitals can occur, such that excitation energy is effectively shared between two molecules in a process known as delocalized exciton coupling and involving electron exchange. This process occurs at faster rates than resonance energy transfer, which is thus not significant at small intermolecular separations.

In an organism such as *R. sphaeroides* the light-harvesting or antennae chlorophylls are associated with two light-harvesting complexes known as LH 1 and LH 2. The former is a protein complex that is closely associated with the reaction centre, whilst LH 2 is located further away from the reaction centres but sufficiently close to LH 1 to permit energy transfer to it (Fig. 6.3). Each of the light-harvesting complexes have two polypeptide chains, known as α and β (which differ somewhat in the two complexes). LH 1 contains 24 molecules of bacteriochlorophyll and 24 carotenoid molecules while each LH 2 has 18 molecules of bacteriochlorophyll and 9 carotenoid molecules (Papiz *et al.*, 1989). The amino acid sequences of the polypeptides, which contain approximately 50 amino acids, indicate that each will form a single transmembrane α-helix.

Recent X-ray diffraction analysis of a light-harvesting complex (LH 2 type) from *Rhodopseudomonas acidophila* at 0.32 nm resolution (Papiz *et al.*, 1989) indicates that the polypeptides are arranged as an $\alpha_6\beta_6$ complex organized in

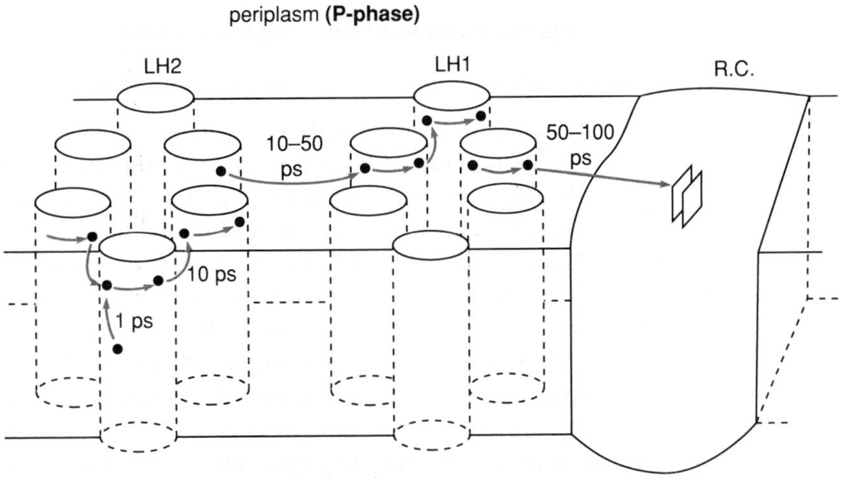

Figure 6.3   The organization of the two light-harvesting complexes, LH 1 and LH 2 (antennae), in the purple non-sulphur photosynthetic bacterium *R. sphaeroides*.
For LH 2 each large cylinder represents an $\alpha_6\beta_6$ polypeptide unit (for which there is good evidence), whereas for LH 1 the cylinder represents an assembly of α and β chains, probably in total $\alpha_{12}\beta_{12}$ per reaction centre. The small filled circles represent bound pigments. Excitation energy first equilibrates within an LH 2 unit (about 1 ps), then with other LH 2 units, and then with LH 1 units (10–50 ps). The excitations are collected within LH 1 on Bchl$_{896}$ (50–100 ps) before transfer to the special pair of bacteriochlorophylls on the reaction centre (R.C.).

three $\alpha_2\beta_2$-units, with each unit comprising four transmembrane α-helices (R. Cogdell and colleagues, unpublished findings). Model building, largely on the basis of amino acid sequences, suggests that the majority of the chlorophylls are located towards the periplasmic side of the membrane, which, as we shall see later (Section 6.2.2), is the location of the pigment that absorbs light in the reaction centre to initiate photochemistry.

From the moment of absorption of light by a component in LH 2 it takes approximately 100 ps for the excitation energy to reach the reaction centre. Unless there prove to be chlorophyll molecules very closely adjacent, random excitational migration within a light-harvesting complex will occur by the resonance transfer mechanism, which also accounts for the transfer from bacteriochlorophyll in LH 2 to LH 1 and from the latter to the reaction centre (Fig. 6.3). In contrast, transfer from carotenoid pigments to bacteriochlorophyll is always by the delocalized exciton coupling mechanism. The excited state lifetime of carotenoids is too short to permit resonance energy transfer; this restriction means that at least part of a carotenoid molecule must be little further than the van der Waals distance from a bacteriochlorophyll. Very rapid transfer of energy between pigments and onwards to the reaction centre is essential if loss of energy by fluorescence or conversion to heat is to be avoided.

6.2.2 The bacterial photosynthetic reaction centre

Reviews   Feher *et al*. 1989, Deisenhofer and Michel, 1991

The only two cytoplasmic membrane proteins for which high-resolution structures have been obtained by X-ray diffraction analysis at the time of writing are both bacterial photosynthetic reaction centres. Although that from *Rhodopseudomonas viridis* was the first and seminal structural determination, we shall mainly discuss what has been learned about photosynthesis from study of the *R. sphaeroides* system, since this has been studied much more extensively at the functional level. Purified reaction centres from *R. sphaeroides* comprise three polypeptide chains, H, L and M, together with four molecules of bacteriochlorophyll (Bchl), two molecules of bacteriopheophytin (Bpheo), two molecules of UQ and one molecule of non-haem iron. It turns out that spectroscopic and biochemical studies on isolated reaction centres correlate in a very satisfying manner with the structure. First we shall review the key findings from the functional studies.

(a) *P$_{870}$ to Bpheo*

Spectroscopic studies of reaction centres revealed that illumination caused a loss of absorbance (bleaching) at 870 nm consistent with the loss of an electron from a component absorbing at this wavelength. This was supported by the finding that ferricyanide ($E_{m,7}$ for $Fe(CN)_6^{3-}/Fe(CN)_6^{4-}$ + 420 mV) caused a similar bleaching in the dark. The component absorbing at 870 nm was termed P_{870} and it was proposed that the absorption of a quantum led (within about a femtosecond) to a transient excited state, P_{870}^*, in which an electron was raised to a higher energy level — increasing the ease with which the electron can be lost. This is the same as saying that the $E_{m,7}$ for P_{870}^+/P_{870}^* is very negative relative to P_{870}^+/P_{870}.

The electron would then be transferred to an acceptor to generate the bleached (oxidized) product, P_{870}^+. It should be noted that the P_{870}^+/P_{870} redox couple has a rather *positive* $E_{m,7}$ (about $+500$ mV) so that it can act as an electron acceptor from the cyclic electron transport system (Fig. 6.2).

Spectroscopic experiments further indicated that P_{870} was a unique Bchl dimer. The oxidized state P_{870}^+ had an ESR spectrum with a linewidth which was consistent with an unpaired electron delocalized over both Bchl rings. Note that in contrast to a haem group, the electron is lost from the tetrapyrrole rings of the Bchl dimer; unlike Fe^{2+}, Mg^{2+} cannot give up an electron. The crystal structure of the reaction centre is consistent with this model, with two closely juxtaposed Bchl molecules (see Plate III).

Rapid excitation of reaction centres with picosecond laser pulses combined with rapid recording of visible absorption spectra at first suggested that the immediate acceptor of the electron lost from P_{870}^+ was a Bpheo molecule. Bpheo is a chlorophyll derivative, where the Mg^{2+} is replaced by two protons. The transfer of an electron from P_{870}^* to Bpheo can be detected in less than 10 ps (Fig. 6.4), and the resulting $(Bchl)_2^+ \ldots (Bpheo)^-$ biradical (often termed P^F) has a characteristic spectrum. Studies of isolated Bpheo in non-polar solvents suggest that its E_m in the reaction centre is about -550 mV, or more than 1 V more negative than P_{870} in its unexcited state (Fig. 6.2). At low temperatures the spectra of the two Bpheo molecules could be resolved and this permitted demonstration that only one Bpheo normally accepted an electron.

The two additional molecules of Bchl were originally thought to be inactive and were termed the voyeur chlorophylls. However, subsequent to the elucidation of the crystal structure (Plate III) which showed that they flanked P_{870}, additional rapid spectroscopic measurements in conjunction with femtosecond flash excitation studies have indicated that one of the voyeur chlorophylls is an intermediate in the passage of electrons from the special pair to Bpheo. Figure 6.4 shows a current view of the timescale of electron transfer in these initial stages.

(b) *Bpheo to UQ*

The biradical P^F (i.e. $P_{870}^+.Bpheo^-$) is highly unstable, and within 200 ps the electron is transferred from $Bpheo^-$ to UQ. There are two UQ binding sites, designated A and B. The addition of a single electron to the UQ at site A results in the formation of the free radical semiquinone anion, $UQ^{\cdot-}$ which we have previously met in the context of the bc_1 complex (Section 5.8). The effective $E_{m,7}$ of the $UQ^{\cdot-}/UQ$ couple is about -180 mV. The electron is further transferred to the second bound quinone at site B. The timescale for these electron transfers is very rapid (Fig. 6.4). We now have a UQ at A and a $UQ^{\cdot-}$ at B. The latter must be stabilized by its binding site within the protein because there is a strong tendency for the radical ion to disproportionate into $UQ + UQH_2$.

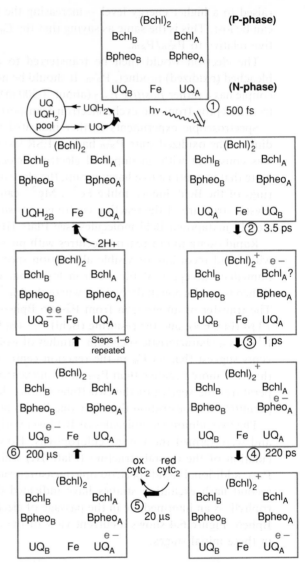

Figure 6.4 Two-electron gating and time-course of electron movement through the bacterial photosynthetic reaction centre from *R. sphaeroides.* For the possible involvement of the bacteriochlorophyll that lies between the special pair and one of the bacteriopheophytin molecules see Holzapfel *et al.* (1990) and Kirmaier and Holten (1991).

(c) *Transfer of the second electron and release of UQH$_2$*

Reduced cyt c_2 is the electron donor responsible for reducing P$_{870}^+$ *in situ*. Since the $E_{m,7}$ of the P$_{870}^+$/P$_{870}$ couple ($+500$ mV) is more positive than that of the cyt c_2 couple ($+340$ mV) the reduction is thermodynamically favoured (Fig. 6.2). A second photon now causes a second electron to pass from P$_{870}$ to the quinone binding site B, once more via a transient UQ$^{\cdot-}$ at A. The UQ at site A thus switches between the oxidized and anionic semiquinone forms and never becomes fully reduced (Fig. 6.4). In contrast, that at B now becomes fully reduced as UQ^{2-}, following which two protons are taken up to give UQH_2 (Fig. 6.4). The UQH_2 is then released to the bulk UQH_2/UQ pool. The two bound UQ molecules thus act as a *two-electron gate*, transducing the one-electron photochemical event into a two-electron transfer. As will be seen in Section 6.3, the protonations of the bound UQ_B play an essential role in the generation of Δp. The cyclic electron transfer is completed by a pathway from the bulk UQ back to cyt c_2; this will be discussed in Section 6.3.

(d)   *Structural correlations*

When the pathway of electron transfer deduced by spectroscopy is correlated with the structure of the reaction centre (Plate III), we see that transfer of electrons from the $(Bchl)_2$ dimer via Bpheo to UQ_A and then UQ_B fits with the spatial distribution of the groups. However, two obvious puzzles present themselves. Firstly there appear to be two branches down which the electron might flow, since the redox carriers are arranged in the reaction centre with close to two-fold symmetry. Yet all the evidence points, including photo-reduction of only one Bpheo and linear dichroism absorption experiments with the knowledge of the crystal structure, to only the right-hand branch (as depicted in Plate III) being significantly active. The reason why the electrons flow more slowly down the left-hand branch is not known, but may be related to subtle yet important differences in the relative orientation of the groups that carry the electrons. Second, no role has so far been found for the non-haem iron which the structure shows lies between the two quinone sites; the iron atom can be removed without unduly affecting the performance of the reaction centre.

The two quinone binding sites have distinct properties. It is clearly important that both sites can stabilize the anionic form of UQ, but at the same time the B site must be suitable for the protonation events. Thus the B site is more polar and there is a pathway which can be discerned for UQ from the bulk phase to enter the B site with the head group coming in first. The B site is not in direct contact with the aqueous phase, suggesting that side groups of the protein may be responsible for transferring protons from the bulk phase. A glutamate residue (212) on the L-chain lies between this site and the aqueous phase. Site-directed mutagenesis of this amino acid drastically attenuates the rate of protonation without affecting the rate of electron transfer to UQ_B, thus implicating the carboxylate side chain in proton transfer (Paddock *et al.*, 1989). The neighbouring aspartate-213 has also been implicated (Takahashi and Wraight, 1992).

X-ray diffraction data for purified reaction centres does not give information on the orientation of the complex in the intact membrane. However, cyt c_2 is located in the periplasm, in common with many other bacterial c-type cytochromes, while the H-subunit was susceptible to proteolytic digestion and recognition by antibodies only in inside-out membrane vesicles (i.e. chromatophores, see Chapter 1). Thus the reaction centre is orientated with the special Bchl pair towards the outside (periplasmic) surface of the bacterial cytoplasmic membrane, and with the two UQ binding sites towards the cytoplasm. The L- and M- chains each have five transmembrane α-helices, while the single α-helix of the H-subunit also spans the membrane (Plate III). Electron transfer takes place through the redox groups bound mainly to the L-subunit. These α-helices contain predominantly hydrophobic amino acids and appear to provide a rigid scaffold for the redox groups. The importance of minimizing relative molecular motion of these groups is illustrated by the finding that the rates of some of the electron-transfer steps from the Bchl special pair to the Q_A site increase with decreasing temperature.

(e) Charge movements

With the orientation shown in Plate III, light will cause an inward movement of negative charge from cyt c_2 to UQ_B, where the translocated electrons meet with protons coming from the cytoplasm. Thus the net effect of both of these charge movements is to transfer negative charge into the cell (i.e. from periplasm (P-phase) to cytoplasm (N-phase)), contributing to the generation of a Δp (positive and relatively acidic outside).

The inward movement of the electron through the reaction centre had already been detected using the carotenoid band shift (see Chapter 4) as an indicator of membrane potential. Using chromatophores, three phases of development of a membrane potential followed a short saturating flash of light. The first corresponded kinetically to the transfer of the electron from P_{870} to UQ_B, the second to the reduction of P_{870}^+ by cyt c_2, whilst the third, and much the slowest, phase corresponded to the return of the electrons from UQH_2 to cyt c_2, and could be blocked by antimycin or myxothiazol. It should be noted that the transfer of electrons between the UQ A and B sites is parallel to the membrane and does not contribute to the establishment of a membrane potential.

The contribution of each electron transfer within the reaction centre to the development of membrane potential is related to the distance moved by the electron perpendicular to the membrane and to the dielectric constant of the surrounding membrane. Thus the movement from cyt c_2 to the special pair would make less contribution than the transfer from Bpheo to Q_A through a more hydrophobic (i.e. low dielectric constant) environment. Finally, the uptake of the protons into the Q_B site contributes about 10% to the overall charge separation across the membrane.

Despite the satisfying correlation between structure and function, it should be appreciated that not everything is known about the functioning of the reaction centre. An essential feature of a reaction centre is that it is not reversible, i.e. the electron must not be allowed to return from $UQ^{\cdot -}$ at site A to reduce P_{870}^+, even though the semiquinone is thermodynamically capable of doing this by virtue of the more negative $E_{m,7}$ of its redox couple with UQ. The kinetic factors that prevent such reversal from $UQ_A^{\cdot -}$ or from any of the other components in the reaction centre to the special pair are not understood. Reversal occurs 10^4 times more slowly than the forward reaction. This is the reason for the almost perfect quantum yield, i.e. one photon results in the creation of one low-potential electron. The cost of this irreversibility is the loss of redox potential as the electron passes from P^*_{870} to the quinones and the dissipation of 30% of the absorbed energy (Section 6.1).

6.2.3 The *R. viridis* reaction centre

The *R. viridis* reaction centre differs in one major respect from that in *R. sphaeroides* by having an additional polypeptide subunit which contains four *c*-type haems (Plate I). The haem nearest the special pair of bacteriochlorophylls (designated P_{960} in this organism because the absorption maximum and exact structure of the bacteriochlorophylls are different than in *R. sphaeroides*) is the immediate donor to the reaction centre. The electron is thus transferred over a distance of approximately 2 nm (see Section 5.4) from this haem ($E_{m,7} = +370$ mV). The other haems have redox potentials of $+10$ mV, $+300$ mV and -60 mV, listed in order of increasing distance from the special pair. It is not known which of the haems accepts electrons from the cyt c_2, but on the basis of redox potential it would be expected to be either the $+300$ mV or $+370$ mV component. The function of the other two haems in the multi-haem subunit is not known but their $E_{m,7}$ values are such that a role in electron transfer between cyt c_2 and the $+370$ mV component is not obvious. There is also no satisfactory explanation as to why this tetrahaem *c*-type cytochrome is dispensable in *R. sphaeroides* and certain other organisms (e.g. *Rhodobacter capsulatus*). The presence of this cytochrome allowed, if it and the menaquinone (see below) were pre-reduced, a form of the reaction centre containing $(Bchl)_2$ and $Bpheo^-$ to accumulate, because $(Bchl)_2^+$ was eventually reduced by the cytochrome and electrons could not pass from $Bpheo^-$ to the quinone. ESR studies gave important evidence that the electron resided on Bpheo.

The other difference in *R. viridis* is that the Q_A site is occupied by menaquinone rather than UQ. The Q_B site was found to be empty in the initial crystals of the reaction centre, consistent with the ability of UQ at this site to dissociate and equilibrate with the bulk pool.

6.3 THE GENERATION BY ILLUMINATION OR RESPIRATION OF Δp IN PURPLE BACTERIA

We have already seen from the structure of the reaction centre that absorption of light causes the movement of negative charge into the cell and that the optical spectral changes in carotenoids (Chapter 4) can be used to follow this and other charge separations (Section 6.2.2e). The slowest of the three phases of development of the carotenoid shift that are observed following exposure of chromatophores to very short saturating flashes of light is blocked by inhibitors of the cyt bc_1 complex (Section 6.2.2e) and therefore must correspond to the movement of charge across the membrane by this complex. Such carotenoid measurements together with measurements of light-dependent proton uptake by chromatophores provided important evidence in favour of the Q-cycle mechanism described for the cyt bc_1 complex in Chapter 5.

Since electrons are retained within the cyclic pathway (Fig. 6.1) while protons are taken up and released, only the latter need be considered when calculating the overall charge movements per cycle. Since, per electron, the reaction centre takes one proton from the cytoplasm while the bc_1 complex takes up one proton from the cytoplasm but releases two protons to the periplasm, two protons are translocated for each electron handled. An implication of this stoichiometry is that if the H^+/ATP ratio is 3 (Chapters 4 and 7) then the $ATP/2e^-$ ratio will be 4/3.

Cyclic electron transfer in *R. sphaeroides* and related organisms generates Δp but does not produce reducing equivalents for biosynthesis. Provision of such reducing equivalents (i.e. NADPH) often requires Δp to drive reversed electron transport as well as ATP synthesis. Thus if, for example, the organism is growing on H_2 and CO_2 then electrons from H_2 will be fed via hydrogenase into the cyclic electron-transport system at the level of ubiquinone and driven by reversed electron transfer through a rotenone-sensitive NADH dehydrogenase (probably analogous to complex I, Section 5.6) to give NADH. Subsequent formation of NADPH, presumably via the transhydrogenase reaction (Section 5.11), then provides the reductant for CO_2 fixation. On the other hand if the organism is growing on malate, which has approximately the same oxidation state as the average cell material, electrons will not be fed into the cyclic system and thus reversed electron transport will not be significant. Sulphide and succinate are further examples of substrates in which electrons are fed in at ubiquinone and there are also electron donors in some organisms that donate to the *c*-type cytochromes. Thus the extent and the location at which electrons from a growth substrate are fed into the cyclic electron transport system is dependent on the nature of the substrate. It is crucial that the cyclic electron-transport system does not become over-reduced. If every component were to be reduced then cyclic electron transport could not occur; the mechanism whereby over-reduction is avoided is not fully understood.

R. sphaeroides, in common with many other photosynthetic bacteria, can grow aerobically in the dark. Oxygen represses the synthesis of bacteriochlorophyll and carotenoids, and so the reaction centre is absent. However, the *b* and *c* cytochromes are retained, and a terminal oxidase is induced, which in the case of *R. sphaeroides* is a Cu-containing protein very similar to the mitochondrial complex IV (Section 5.9). By using the preexisting cytochromes, the bacterium can therefore switch very economically between anaerobic photosynthetic growth and aerobic growth in the dark by assembling respiratory and photosynthetic chains with common components (Fig. 6.1).

Cyt c_2 has long been regarded as an essential component in the cyclic electron-transport system. However, recent gene deletion experiments with both *R. sphaeroides* and *R. capsulatus* have shown that cyt c_2 is dispensable for photosynthesis because an alternative *c*-type cytochrome can substitute. The implications of this finding are not fully understood, but it is still thought that in wild-type cells cyt c_2 plays a major role in cyclic electron transport. Organisms such as *R. sphaeroides* can also use certain anaerobic electron acceptors, including at least some of the oxides of nitrogen (Ferguson *et al.*, 1987). Their electron-transport chains are thus even more versatile than Fig. 6.1 suggests because they can also incorporate some of the components shown for *P. denitrificans* in Section 5.13.1.

6.4 THE ELECTRON-TRANSFER AND LIGHT-CAPTURE PATHWAY IN GREEN PLANTS AND ALGAE

Photosynthetic electron transfer in chloroplasts has two features not found in purple bacteria: (1) it can be non-cyclic, resulting in a stoichiometric oxidation of H_2O and reduction of $NADP^+$; (2) two independent light reactions act in series to encompass the redox span from $H_2O/\frac{1}{2}O_2$ to $NADP^+/NADPH$ (Fig. 6.5).

The presence of two reaction centres was indicated by a classical observation known as the *red drop*. Algae illuminated with light in the range from 400–680 nm very effectively evolved oxygen. However, if light with a wavelength greater than 680 nm was used then the efficiency fell very sharply and light >690 nm was essentially ineffective. This in itself merely showed that there was a component which required light <690 nm, either through direct illumination or by energy transfer from shorter wavelengths. What was more striking was that the oxygen evolution produced by relatively weak, non-saturating, light at 650 nm could be increased by simultaneous illumination at 700 nm (the enhancement effect). The interpretation was that two photosystems were involved. One had an absorbance centred at 700 nm, which could be supplied either by direct illumination with 700 nm light or by energy transfer from pigments in the antennae that absorbed the 650 nm light. The second photosystem, with an absorbance maximum at 680 nm, could be

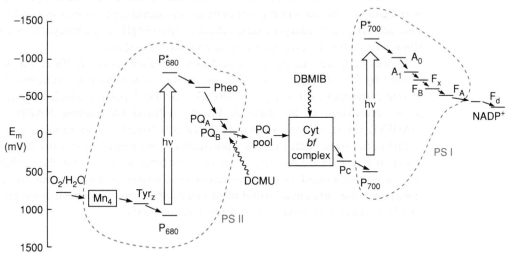

Figure 6.5 Two light-driven reactions, catalysed by PS I and PS II, operate in series in thylakoid membranes to drive electrons from water to $NADP^+$.
The Z-scheme representation is shown with excited state redox potentials where appropriate. The sites of action of DCMU and DBMIB are shown schematically by a wavy arrow. Pc is plastocyanin. Details of the components within the two photosystems (PS I and PS II) are given in Figs 6.6 and 6.8.

excited by wavelengths up to that value, but not by the lower energy 700-nm light. In a development of this experiment it was shown that if the two light beams were applied separately, there was a transient release of oxygen when the shorter wavelength was applied after the 700 nm light (Gregory, 1989).

The two photosystems are now known to be arranged in series (Fig. 6.5). The system that requires 680 nm light for excitation is known as *photosystem II*, PS II, and serves to abstract electrons from water and raise them to a sufficiently negative potential to reduce plastoquinone (PQ) to plastoquinol (PQH$_2$), which are respectively very similar to UQ and UQH$_2$ in structure, (Fig. 6.5) and function. Thus PQH$_2$ acts as the reductant for the cytochrome *bf* (formerly called *b$_6$f*) complex which is in turn very similar to the *bc$_1$* complex of respiratory chains. Accordingly, *bf* possesses two *b*-type haems attached to a single polypeptide chain, a high-potential Fe/S protein and a *c*-type cytochrome, cyt *f*, with redox potential and topology very similar to those of cyt *c$_1$* (the designation *f* is for historical reasons).

Whereas the mitochondrial *bc$_1$* complex acts as an electron donor to the peripheral cyt *c*, the *bf* complex passes electrons to *plastocyanin* which is a peripheral protein located at the lumenal side of the thylakoid membrane. The redox centre in plastocyanin is a Cu ion which undergoes a 1e$^-$ oxidation/reduction reaction. The environment of the Cu within the protein is such that its $E_{m,7}$ is $+370$ mV, very different from what it would be in aqueous solution. This, together with the characteristic blue absorbance spectrum and e.p.r. spectrum is diagnostic of a *type I* Cu centre. X-ray diffraction studies shows the Cu to have a highly distorted tetrahedral co-ordination geometry with ligands from the sulphurs of cysteine and methionine as well as two histidine side chains. In energetic terms the $E_{m,7}$ span PQH$_2$ to plastocyanin is very similar to UQH$_2$ to cytochrome *c* in respiratory systems.

Plastocyanin acts as the electron donor to *Photosystem I*, PS I, which is excited by 700 nm light. PSI raises the energy of the electron sufficiently to enable *ferredoxin*, Fd, ($E_{m,7} = -530$ mV), an Fe/S protein which undergoes a 1e$^-$ reduction, to be reduced. Finally reduced Fd reduces NADP$^+$ to NADPH, via an enzyme known as ferredoxin-NADP$^+$ oxidoreductase.

Figure 6.5 shows that the redox potential of the electron as it is driven from water to NADPH follows the shape of a distorted letter N with the uprights significantly displaced: this scheme has become known as the *z-scheme* because it was once presented in a format displaced by 90° from the present convention in which redox potential is shown on a vertical axis.

6.4.1 Plant antenna systems

Just as in bacterial photosynthesis, light harvesting or antennae complexes are required in the thylakoid membrane. There are distinct polypeptide complexes, associated with the two photosystems I and II, with non-covalently attached chlorophylls (two types *a* and *b*) and xanthophyll, known as LHC 1 and LHC II. A principal polypeptide in LHC II is known as LHC IIb (or sometimes called chlorophyll binding protein). This has a molecular weight of 25 K and associated with it are eight molecules of chlorophyll a, seven of chlorophyll b and carotenoids. The electron microscopy/diffraction approach first applied to bacteriorhodopsin (see later) has been successfully applied to this protein and a 6 Å (0.6 nm) structure obtained (Kuhlbrandt and Da Neng Wang, 1991). This shows that three polypeptide chains associate to form trimers. There are three transmembrane helices within each polypeptide (Plate II); a striking feature is that two of these helices are longer (46 and 49 amino acids) than has been observed previously for transmembrane helices in the three other membrane proteins of known structure, a bacterial antennae complex, the reaction centre and bacteriorhodopsin (Sections 6.2.1, 6.2.2 and 6.5). The structure is not closely related to any other membrane protein. The electron density allows fifteen chlorophyll molecules to be discerned (Plate II) lying on the outside of the helical structure which is perceived as a scaffold for the pigments. Energy transfer between chlorophylls attached to a particular polypeptide chain, from chlorophylls on one chain to another, as well as from LHC IIb to a photosystem will mainly be via the resonance energy transfer mechanism, although some chlorophylls within a monomeric unit may be sufficiently closely juxtaposed and appropriately oriented to involve delocalized exciton-coupling (Section 6.2.1).

6.4.2 Photosystem II

Reviews   Rutherford 1989, Andersson and Styring 1991

The green plant photosystems have been more difficult to purify and characterize than the single reaction centre of the purple bacteria, and consequently less detailed information is available. One difficulty is that less than 1% of the pigments are involved in the photoreaction, with the remainder acting as antennae. Nevertheless, the sequences of many of the polypeptides associated with PSII have been obtained and two of them, known as D_1 and D_2, have sequence similarity, particularly in respect of the number of transmembrane spanning regions and predicted pigment binding sites, with the L and M subunits respectively of the bacterial reaction centre. An important distinction is that the postulated (much, but not all, the spectroscopic evidence supports it) special pair chlorophyll (P_{680}) in PSII is reduced by an electron originating from water rather than from cytochrome c_2. Figure 6.6 shows a postulated organization for PSII.

The water splitting reaction is, together with its opposite number the terminal oxidase of the respiratory chain, one of the most intriguing reactions in bioenergetics. In air the E_h for the $O_2/2H_2O$ couple is $+810$ mV (Section 5.9), so to abstract electrons from water requires a redox centre which is even more electropositive and capable of reacting spontaneously with water! It is established that the water-splitting component contains 4 manganese atoms, and that four sequential events are required to abstract the $4e^-$ from 2 molecules of water to yield O_2 and liberate $4 H^+$. A possible mechanism, involving the

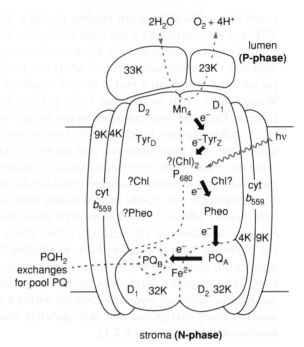

$2H_2O \quad O_2 + 4H^+$

lumen
(P-phase)

33K   23K

D_2 Mn_4 D_1

9K 4K Tyr_D e^--Tyr_Z $h\nu$

?$(Chl)_2$

?Chl P_{680} Chl?

cyt b_{559} ?Pheo e^- Pheo   cyt b_{559}

e^- 4K 9K

PQH_2 exchanges for pool PQ PQ_B e^- PQ_A

Fe^{2+}

D_1 32K D_2 32K

stroma **(N-phase)**

Figure 6.6 Possible organization of polypeptides and cofactors in photosystem II.
Model based on analogy with bacterial reaction centre (i.e. assumed two-fold symmetry with only one active branch from a special pair of chlorophylls (P_{680}) to plastoquinone at Q_A; two chlorophylls close to the special pair and a diffusible plastoquinol from the Q_B site). The role of pheophytin (Pheo) is generally accepted. Several polypeptides have been identified and these are indicated with their molecular weights. Tyr_D is a second tyrosine residue that can be photooxidized but which is not on the forward electron-transport pathway under normal conditions. Based on Rutherford, (1989).

four oxidation states (traditionally termed S states) commonly described for the water splitting reaction is given in Fig. 6.7.

The electrons from the water splitting reaction are not transferred to the P_{680} directly, but rather via a specific tyrosine residue. Loss of an electron to P_{680} from this tyrosine generates a radical that is a neutral species since a proton is also lost, probably to a neighbouring histidine residue. The tyrosine residue, named Z^+ before its molecular identification, partly through study the properties of a photosystem II in which the critical tyrosine had been changed to phenylalanine by site directed mutagenesis, in turn regains an electron from the water-splitting reaction.

The pathway of electron transfer from P_{680} to PQ is postulated to be closely analogous to the corresponding reactions in the bacterial reaction centre, including the presence of two sequential quinone binding sites A and B. Figure 6.6 summarizes these points, some of which have experimental support but all of which cannot be regarded as proven. The Q_B site is thought to be the site

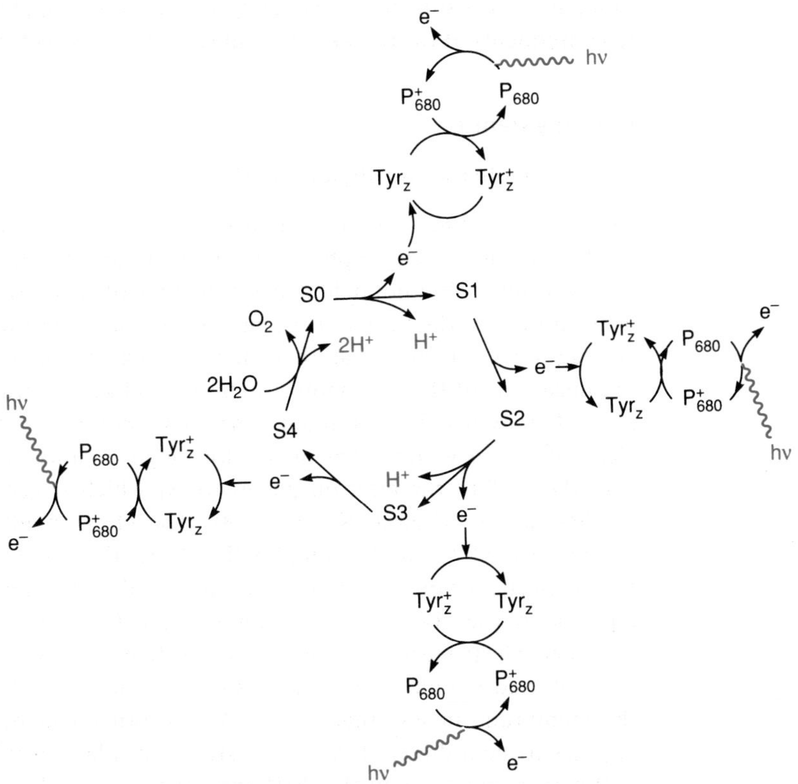

Figure 6.7 The water-splitting reaction of photosystem II.
Four quanta are required to abstract four electrons from two H_2O. In the
dark the water-splitting centre is in state S1. The steps at which H_2O binds
are not known but the H^+ and O_2 release steps are thought to be as shown.
The S4 state has four positive charges due to the transfer of four electrons
into the photosystem. The water-splitting centre contains Mn, but it is not
clear how many electrons originate from the Mn. Ca^{2+} is required for the
transition from S3 to S4. Removal of Ca^{2+} traps the S3 state, in which
ESR indicates that an electron has been lost from a histidine adjacent to
a Mn atom (Boussac *et al.* 1990).

of action of the inhibitor DCMU which blocks the activity of PS II. The similarity to the bacterial reaction centre is strengthened by the finding that a
mutant of the bacterial reaction centre has been identified which is sensitive
to DCMU, unlike the wild-type, and in which the e.p.r. spectrum of $UQ^{\cdot -}$ has
similarities with that of $PQ^{\cdot -}$ bound to PS II.

PSII also contains two molecules of the *b*-type cytochrome known as b_{559}.
The role of the cytochrome is not known for certain, but it may be capable
of reducing the oxidized P_{680}^+ under certain conditions and allowing a cyclic
pathway of electron flow from PQH_2 at the Q_B site. This apparently wasteful
pathway (which would not generate Δp) may arise under conditions of high
light and temperature when electron flow from the water splitting centre to
P_{680}^+ might be inadequate to reduce the latter sufficiently rapidly. P_{680}^+ is

thought to be a sufficiently oxidizing species to be capable of causing damage to components of the thylakoid membrane if it persists for a significant period.

6.4.3 Photosystem I

Review   Golbeck and Bryant, 1991

The understanding of photosystem I (PSI) is less securely based than that of PSII. PSI contains chlorophyll as the pigment absorbing at 700 nm (P_{700}), and it very probably but not definitely (at the time of writing) functions as a dimer, as is known for the bacterial special pair and as is postulated for PSII. There is suggestive evidence that there may be significant analogies between the organizations of the two photosystems (Nitschke and Rutherford, 1991). The $E_{m,7}$ of unexcited P_{700} is about $+450$ mV. Following excitation of P_{700} the electron arrives within 10 ps on a chlorophyll a molecule known as A_0. After a further 100 ps the electron passes to A_1 which is generally believed to be *phylloquinone*, vitamin K_1 (Fig 6.8). The electron ultimately reaches FeS centres located within the complex (Fig. 6.8). Presumably one of these serves as the electron donor to reduce the iron–sulphur protein ferredoxin in the aqueous phase. Bound ferredoxin has an $E_{m,7}$ of -530 mV, and so is extremely electronegative. The carotenoid band shift (Section 6.3), which was first detected in chloroplasts, indicates that the electron is transferred across the membrane in less than 20 ns. At the other end of the complex, there appears to be a direct electron transfer from plastocyanin to P_{700}^+.

Whereas photosystem II (PSII) appears to be closely related to the reaction centre of *R. sphaeroides* and *R. viridis*, there are grounds for supposing that the reaction centre in green sulphur bacteria and heliobacteria may be closely related to photosystem I (PSI) (Nitschke and Rutherford, 1991). An important difference between the reaction centres in at least some green bacteria and purple bacteria is that photochemistry in the former produces a species with a much more reducing, i.e. negative, E_m value, to the extent that in some species $NAD(P)^+$ can be reduced directly. Thus in these green bacteria electrons from donors such as sulphide can be added at quinone and exit from reaction centre to a component, probably a ferredoxin, that can reduce $NAD(P)^+$ without any requirement for Δp-driven reversed electron transport (see Gregory, 1989).

6.4.4 Δp generation by the Z-scheme

A carotenoid shift response indicates that both the photosystems are oriented across the membrane (Figs. 6.6 and 6.8). Further evidence for the orientation of PS II comes from the observation that the protons liberated in the cleavage of H_2O are initially released into the lumen, indicating that oxidation of water occurs on the inner, i.e. lumenal or P, side of the membrane. Also, a radical anion form of plastoquinone bound to the reaction centre must be located close to the N-side of the membrane, since it can be made accessible to

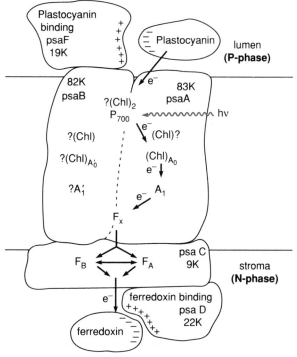

Figure 6.8   A plausible model for Photosystem I.
The model is heavily based on analogy to the bacterial reaction centre. There is no definite proof (but some evidence is in favour) for a special pair of chlorophylls as P_{700} nor for the adjacent monomeric chlorophylls ((Chl)?). A_0 and A_1 are widely considered to be chlorophyll and phylloquinone; F_X, F_A and F_B are iron–sulphur centres. Some of the polypeptides are shown schematically with their molecular weights and gene names. Based on Nitschke and Rutherford (1991) and Golbeck and Bryant (1991).

impermeant electron acceptors such as ferricyanide after brief trypsin treatment. Ferredoxin and ferredoxin–NADP$^+$ reductase are accessible to added antibodies, whereas plastocyanin is not. These observations all suggest that PSI is oriented across the membrane as shown in Fig. 6.8.

The translocation of each electron from water to NADP$^+$ through the two photosystems is equivalent to the translocation of two positive charges into the lumen (Fig. 6.9). In addition to this, proton translocation can be associated with the *bf* complex. If this were to function analogously to the closely related *bc₁* complexes of mitochondria and purple bacteria it would be predicted that four protons would appear in the lumen for each pair of electrons flowing from Q_B to plastocyanin. At high light intensities some experimental observations suggest that only 2 H$^+$ per 2 e$^-$ are released, in which case the complex would not contribute any movement of charge across the membrane (Fig. 6.9). If this is truly the case, and more experiments are needed for confirmation, then the *bf* complex must function differently from the *bc₁* complex. Definite

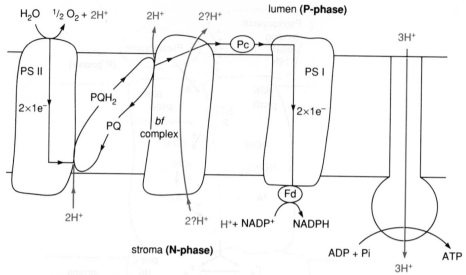

Figure 6.9 Stoichiometry of charge translocation and generation of Δp associated with electron transfer from water to $NADP^+$.
The movement of the electron through the two photosystems moves negative charge from the P- to the N-phase. The other passage of the electron across the membrane, from plastoquinol to plastocyanin will, in mechanism A not result in the movement of any charge if the bf complex releases two protons into the P-phase, whereas in mechanism B release of four protons to the P-phase, and uptake of two from the N-phase, moves positive charge into the P-phase (see text). In mechanism A four positive charges reach the P-phase for each two electrons passing from H_2O to $NADP^+$, four protons are released into the P-phase but only three protons are taken from the N-phase. This is because reduction of $NADP^+$ requires two electrons but one proton. In mechanism B six positive charges reach the P-phase, six protons are released and five protons taken up per 2 e^-. P/2e^- ratios depend on the H^+/ATP ratio, shown here as 3.

differences include an insensitivity of the bf complex to both antimycin and myxothiazol. 2,5-Dibromo-3-methyl-6-isopropylbenzoquinone (DBMIB) is an inhibitor of the cytochrome bf complex which acts at the Q_p site (i.e. on the lumenal side of the membrane) and is thus equivalent to the locus of action of myxothiazol on the cyt bc_1 complex (Section 5.8.1).

The bf complex has the two b-type haems characteristic of both the mitochondrial and purple bacterial bc_1 complexes. It has been proposed (Cramer and Knaff, 1989) that they do not differ in $E_{m,7}$, unlike their counterparts in the cyt bc_1 complexes, where the haem on the N-side of the membrane has an $E_{m,7}$ some 150 mV more positive than the other. It was proposed in Section 5.8.4 that this difference in $E_{m,7}$ allowed the electron to pass across the membrane against a 150-mV membrane potential. Since thylakoids express their Δp almost entirely as a ΔpH (Sections 4.2.4 and 6.4.7) no difference in $E_{m,7}$ would be needed. However, the notion of identical $E_{m,7}$ values for the two

P-phase

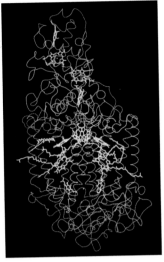

N-phase

Plate I   The three-dimensional structure of the *Rhodopseudomonas viridis* reaction centre.
The tetrahaem cytochrome *c*, green; M subunit, blue; L subunit, brown; H subunit, purple; redox centres, yellow. The Q_A site (towards bottom on right) is occupied by menaquinone. Reproduced with permission from Deisenhofer and Michel (1989) *EMBO J.* **8**, 2154.

stroma **(N-phase)**

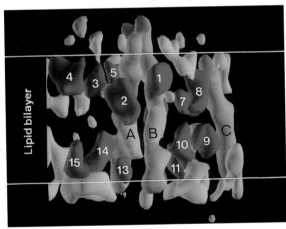

lumen **(P-phase)**

Plate II   The structure of a light-harvesting complex from thylakoids, showing the disposition of the chlorophylls.
Side view of the thylakoid LHC II monomer. The three transmembrane helices (purple) are labelled A, B and C. The order, starting at the N-terminus, is probably B, C, A. Thirteen of the 15 chlorophylls (green) are shown (chlorophylls 6 and 12 are hidden behind helix B). Helix C is perpendicular to the membrane plane and fully embedded in the bilayer, helices A and B are longer, are inclined at 30° to the membrane normal and protrude into the stroma (N-phase). Electron density from neighbouring units is in pale blue. Reproduced with permission from Kuhlbrandt and Da Neng Wang (1991).

P-phase

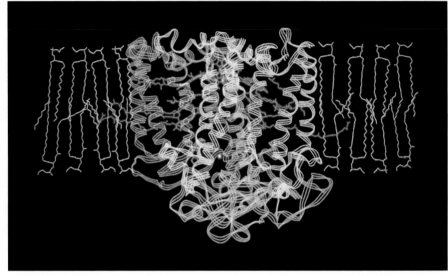

N-phase

P-phase

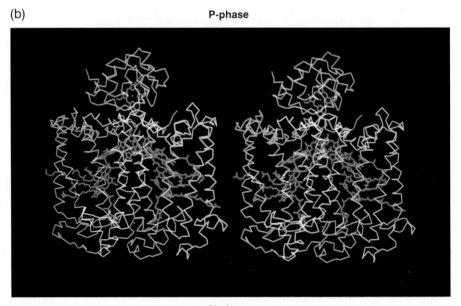

N-phase

(c)

P-phase

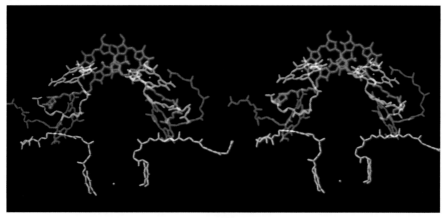

N-phase

(d)

P-phase

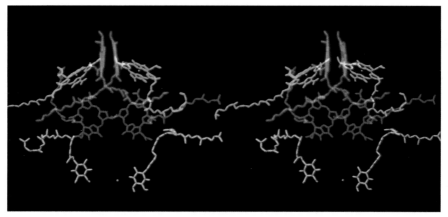

N-phase

Plate III   Structural features of the photosynthetic reaction centre from *R. sphaeroides*. (a) The estimated position of the reaction centre in the lipid bilayer showing a schematic model arrangement of phosphatidyl ethanolamine as sole lipid. Colour code: L subunit, yellow; M subunit, blue; H subunit, green; cofactors, red; carotenoid, purple. The dot near the cytoplasmic or N-phase is the Fe atom that lies on the two-fold symmetry axis that is in the plane of the paper. The right-hand branch of cofactors, mainly associated with the L chain, is photochemically active. (b) A stereoview of a possible docking mode of cytochrome c_2 (green with its haem in red) with the L (yellow) and M (blue) chains. Their associated cofactors are shown in red; the H-chain has been omitted for simplicity. (c and d) Stereoplots of the cofactors. These are related to one another by a $90°$ rotation around the two-fold symmetry axis. Colour code: bacteriochlorophyll dimer, red; bacteriochlorophyll monomers, green; bacteriopheophytins blue; ubiquinones (Q_a at right-hand side), yellow; Fe atom (dot between two ubiquinone sites); yellow. Reproduced with permission from: (a) Feher *et al.* (1989); (b–d) Allen *et al.* (1987) *Proc. Natl. Acad. Sci. USA*, **84**, 6165 and 5733.

P-phase

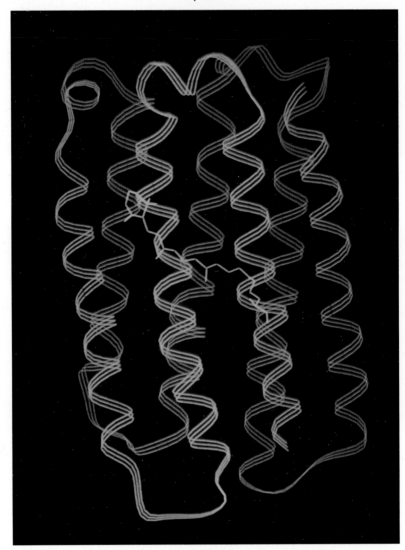

N-phase

Plate IV   The structure of bacteriorhodopsin.
The ribbon diagram shows the backbone of the polypeptide with its seven
α-helices. The retinal is shown attached (via a Schiff base) on lysine 216
of helix G. Reproduced from Henderson *et al.* (1990) with permission.

b-type haems in *bf* is not accepted by many investigators who propose that the values differ just as they do in the bc_1 complex.

Figure 6.9 shows the overall proton movements occurring in non-cyclic electron transport in the thylakoid. The overall stoichiometry is either $4H^+/2e^-$ or $6H^+/2e^-$ delivered to the lumen, depending on the value adopted for the *bf* complex. If three protons must be translocated through the thylakoid ATP synthase to synthesize one ATP, then depending on the $H^+/2e^-$ ratio we expect that either 1.3 or 2 ATP are synthesized for each NADPH synthesized.

6.4.5 Cyclic electron transport

The main destination of the NADPH produced by non-cyclic electron flow is the Calvin cycle, which fixes CO_2 in an overall process which requires three ATP for each two NADPH. Thus the stoichiometry of ATP synthesis associated with electron flow leaves a shortfall if the $H^+/2e^-$ ratio is only 4. One mechanism in which oxygen could act as an electron acceptor to make good this shortfall is described in Section 6.4.6, but the possible shortfall can also be counteracted by cyclic electron transport – which occurs when electrons are able to return from ferredoxin to PQ (Fig. 6.10). Cyclic electron transport can occur (but like all cyclic processes is difficult to observe in the steady state) when thylakoids are illuminated with 700-nm light such that only PS I is active; under these conditions ATP can still be synthesized. Furthermore, there are cells in which only PS I appears to be active and cyclic electron transport occurs. These include heterocysts of cyanobacteria which fix nitrogen and thus require an anaerobic environment, and the bundle-sheath cells of some C_4 plants. However, measurement of quantum efficiencies of Photosystems I and II in leaves at ambient CO_2 suggests that cyclic phosphorylation is in general a minor contributor, although may be important during an induction phase when leaves are illuminated.

Although the idea of cyclic electron flow has been accepted for some time the pathway is surprisingly not yet understood. A plausible scheme is shown in Fig. 6.10 , involving a ferredoxin/PQ oxidoreductase. Such an enzyme and its constituent polypeptides have not been firmly identified (at the time of writing some researchers prefer a scheme in which electrons are passed from

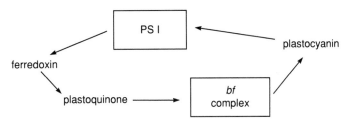

Figure 6.10 Physiological cyclic electron transport in thylakoids —a plausible scheme, assuming the existence of ferredoxin-plastoquinone oxidoreductase.

ferredoxin to the cyt *bf* complex). It is possible that there is involvement of polypeptides with sequence similarities to components of mitochondrial NADH dehydrogenase (complex I) that are predicted from gene sequencing to occur in thylakoids. Recall that mitochondrial complex I is also a quinone reductase (there is no convincing evidence, or rational role, for an NADH dehydrogenase in thylakoids of green plants). Whatever the uncertainty surrounding the molecular components involved in transfer of electrons in cyclic electron transport from ferredoxin to PQ, it is known that antimycin inhibits the process. The locus of this inhibition is not the cyt *bf* complex (see Section 6.4.4). Figures 6.9 and 6.10 together show that, depending on the $H^+/2e^-$ stoichiometry of the *bf* complex, each turn of the cycle would result in two or four protons translocated per 2 e^-. The latter stoichiometry is the same as for cyclic electron transport in purple bacteria.

What regulates the relative activities of the two photosystems and thus of cyclic and non-cyclic electron transport? Some clue may come from the arrangement of the photosystems in the thylakoid membrane. PS II can be found in the stacked regions of the thylakoids, whereas PS I, which has to deal with the large ferredoxin substrate, is restricted to the unstacked regions (Figs 1.6 and 6.11). Since the light-harvesting complexes transfer energy to

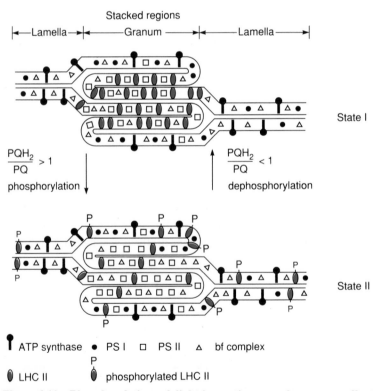

Figure 6.11 Phosphorylation of light-harvesting complexes may affect their distribution between stacked (rich in PS II) and unstacked (rich in PS I) regions of the thylakoid membrane.

the photosystems by resonance energy transfer (Section 6.2.1), the effectiveness of which decreases as the sixth power of the distance, the separation between light-harvesting complexes and photosystems will be critical. Light-harvesting complexes can be phosphorylated, and this is thought to cause them to be excluded from the stacked regions, thus decreasing the energy transferred to PS II. The extent of phosphorylation increases as the ratio PQH_2/PQ increases (implying that PS II is becoming more active than PS I), directing the light-harvesting complexes towards PS I in the non-stacked regions (Allen, 1992) and restoring the balance by supplementing other antennae for PS I (Fig. 6.11). Such a role for phosphorylation has not, however, been indisputably established at the time of writing.

The condition to which chloroplasts revert in the dark, in which LH II is predominantly associated with PS II, is often known as state I, whilst state II refers to the situation in which at least some of the LH II is considered to have migrated to the stroma lamellae region enriched in PS I. The terminology of states I and II is not to be confused with the description of mitochondrial respiratory states (Chapter 4). In the case of the thylakoid, state I was originally defined as the condition in which PS I was over-excited (in which case the thylakoid showed a relatively high fluorescence) and state II the condition where PS II received excess excitation and fluorescence was relatively low. Changes of state are thus induced by absorption of excess excitational energy by one of the two photosystems; the changes occur reversibly over several minutes. State I correlates with the operation of PS II, whilst state II favours cyclic electron flow around PS I. As cyclic electron transport generates ATP but not NADPH, it is often proposed that an ATP requirement of a photosynthetic cell may be involved in controlling the transition to state II.

6.4.6 Other electron donors and acceptors

The non-physiological electron acceptor ferricyanide allows the Hill reaction, a light-dependent oxygen evolution in the absence of $NADP^+$, to be observed. The $E_{m,7}$ of the $Fe(CN)_6^{4-}/Fe(CN)_6^{3-}$ couple, $+420$ mV, is sufficiently positive to accept electrons from reduced plastocyanin, but, since plastocyanin is on the lumenal side of the ferricyanide-impermeable membrane, the couple accepts electrons from a donor on the stromal side of the PS I complex. Thus oxygen evolution with ferricyanide as acceptor requires the operation of both photosystems. However, if in addition to ferricyanide, a benzoquinone is also present, electrons can be transferred across the membrane from bulk phase plastoquinol (or possibly the Q_B site) to the external ferricyanide without the involvement of PS I (Fig. 6.12).

Operation of PS I in isolation can be achieved by donating electrons from ascorbate via DCPIP to plastocyanin. Illumination will then drive electrons through PS I to $NADP^+$; alternatively $NADP^+$ can be replaced by the oxidized form of the non-physiological acceptor methyl viologen (Fig. 6.12). As would be predicted, either of these reactions can occur in the presence of the

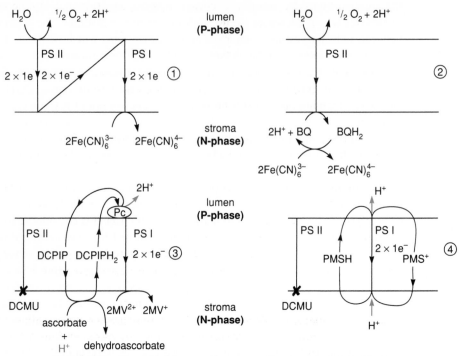

Figure 6.12 Use of redox mediators to dissect the thylakoid electron transport system.

Reaction I is the classic Hill reaction.

PS II inhibitor, DCMU, and can be driven by light of 700 nm which, as explained earlier (Section 6.4), is incompetent to activate PS II. Figure 6.12 also shows how another non-physiological electron carrier, phenazinemethosulphate (PMS), can allow cyclic electron transfer around PS I to occur. Because the reduced form of PMS carries two electrons plus a proton, this form of cyclic electron transport generates a Δp and hence ATP synthesis can be observed.

Thylakoids also catalyse a process known as pseudocyclic electron transport and phosphorylation. Reduced ferredoxin (Fd_{red}) can react with oxygen to give superoxide:

$$4Fd_{red} + 4O_2 \rightarrow 4Fd_{ox} + 4O_2^{\cdot-}$$

The subsequent activity of superoxide dismutase and added catalase in an *in vitro* experiment results in the formation of 3 O_2:

$$4O_2^{\cdot-} + 4H^+ \rightarrow 2H_2O_2 + 2O_2 \quad \text{(superoxide dismutase)}$$

$$2H_2O_2 \rightarrow 2H_2O + O_2 \quad \text{(catalase)}$$

such that the overall reaction becomes:

$$4Fd_{red} + 4H^+ + O_2 \rightarrow 4Fd_{ox} + 2H_2O$$

If these reactions are compared with that for the normal electron flow through both photosystems from H_2O to Fd_{ox}:

$$4Fd_{ox} + 2H_2O \rightarrow 4Fd_{red} + 4H^+ + O_2$$

it is apparent that the cycling of Fd between these two reactions results in no net oxygen consumption or formation, allowing a pseudocyclic electron flow to proceed through both photosystems, with the generation of a Δp and ATP synthesis.

Since the intact chloroplast does not contain catalase, H_2O_2 being instead reduced by ascorbate, when Fd is oxidized *in vivo* by O_2 then one molecule of O_2 will be consumed for each four electrons originating from water and passing to Fd. This overall reaction will lead to ATP synthesis without formation of NADPH and is believed to occur in the intact cell as an alternative to cyclic electron transport to make good any shortfall in ATP synthesis (Section 6.4.5).

In the absence of Fd a similar reaction can be observed if methyl viologen is added, since this dye also reacts with oxygen to give superoxide:

$$MV^+ + O_2 \rightarrow MV^{2+} + O_2^{\cdot -}$$

If, however, superoxide dismutase and/or catalase are absent or inhibited then net O_2 consumption will be observed because formation of four MV^+ produces one O_2 whilst oxidation of four MV^+ consumes four O_2. This is an example of the *Mehler* reaction. A variant is when DCMU is present to inhibit PS II and electrons are donated to PS I from ascorbate via DCPIP and plastocyanin. A light-dependent oxygen consumption is observed driven by the operation of PS I alone.

6.4.7 The proton circuit

The steady-state Δp in thylakoids is, as discussed earlier (Section 4.2.4), present almost exclusively as a ΔpH due to the permeability of the thylakoid membrane to Mg^{2+} and Cl^-. One important consequence of this is that electron transport can be uncoupled from ATP synthesis by ammonium ions or other weak bases. Ammonia enters as NH_3, increasing the internal pH as the ammonium cation is formed. Additional Cl^- uptake occurs in response to further proton pumping with the result that a massive accumulation of NH_4Cl occurs and the thylakoids burst.

Both the photosystems and, debatably, the *bf* complex (Section 6.4.3) contribute to the net translocation of protons across the membrane (Fig. 6.9). In the steady state, ΔpH can exceed 3 pH units, estimated from the accumulation of radiolabelled amines or the quenching of 9-aminoacridine fluorescence. The transient $\Delta\psi$ decays too rapidly to be measured by radiolabelled anion distribution, but can be followed from the decay of the carotenoid shift following single and flash activated turnover of the photosystems. The timescale of the electron-transfer reactions under these conditions is much shorter than for ion movements.

The initial proton movements upon illumination can be followed spectrophotometrically by pH-sensitive dyes such as Neutral Red. Although the transfer of an electron from PS II to PQ is complete within 2 ms, the time-constant for the disappearance of protons from the medium is 60 ms, suggesting that there is an appreciable diffusion-limited barrier for protons on the outer face of the thylakoid membrane.

The chloroplast ATP synthase (Section 7.2) is essentially the sole consumer of Δp in the thylakoid. ATP-dependent proton uptake can be observed in the dark following activation of the latent enzyme by an imposed Δp (Section 7.2). It is important that in the dark this ATP synthase does not wastefully hydrolyse ATP (Chapter 7).

6.5 BACTERIORHODOPSIN AND THE PURPLE MEMBRANE OF HALOBACTERIA

Reviews   Mathies *et al*. 1991, Oesterhelt and Tittor 1989, Khorana, 1988

Halobacteria are extreme halophiles, requiring very high concentrations of NaCl and Mg^{2+} salts for growth, and they can therefore colonize environments such as salt lakes. *Halobacterium halobium* has been most studied. When grown under aerobic conditions the cells utilize a conventional respiratory chain but when growing in the light under conditions of low oxygen tension they synthesize purple patches on their membrane which may be isolated by decreasing the osmolarity of the medium; the purple patches remain intact while the remainder of the cell membrane disintegrates. This 'purple membrane' consists of flat sheets containing a hexagonal crystalline array of a single protein, bacteriorhodopsin, which makes up about 75% of the membrane dry weight, the remainder being phospholipid.

Although *H. halobium* lacks chlorophyll, the cells can use light to translocate protons outwards and synthesize ATP. Photoautotrophic growth is possible but the photochemical events themselves do not generate a reductant. It can easily be shown that proton translocation is due to bacteriorhodopsin, since the purple membranes can be reconstituted into closed vesicles by the addition of phospholipid. These vesicles, which are inverted with respect to the intact cell, take up protons on illumination, and if co-reconstituted with beef heart ATP synthase demonstrate light-dependent ATP synthesis. The importance of this demonstration was that it was difficult to explain in terms of any direct coupling mechanism, since bacteriorhodopsin and beef heart ATP synthase had never met until that moment and so were unlikely to be able to be capable of direct interaction.

The present state of knowledge of bacteriorhodopsin is due to the combination of spectroscopic techniques to monitor changes in the retinal chromophore, detailed structural information and information from site-directed mutagenesis. Although a full high-resolution crystal structure of bacteriorhodopsin has yet to be obtained, considerable structural information is available at a resolution of 0.35 nm in the plane of the membrane, with lower resolution (1 nm) perpendicular to the membrane, by the use of electron cryo-microscopy of tilted samples, exploiting the ability of the protein to form regular two-dimensional arrays in the purple patches. As early as 1975 this technique indicated at 0.7 nm resolution that bacteriorhodopsin had seven membrane-spanning α-helices. The subsequent sequencing of the protein and the improvement of the resolution of the electron microscopic technique has enabled the structure shown in Plate IV to be determined (Henderson *et al.*, 1990). Orientation was determined by labelling from the two sides of the membrane and selective proteolysis which digested the C-terminus only in inside-out vesicles, consistent with a cytoplasmic (N-phase) location of this terminus.

The structure of bacteriorhodopsin is very different from that of the reaction centres discussed above. Although it also transduces light energy into a Δp it differs from other light-driven or respiratory proton pumps in that H^+ translocation is not associated with electron transfer. It provides an insight into a mechanism by which *protons* (as opposed to the electrons translocated by the reaction centres) can be pumped across a membrane. The protein consists of a single 26-kDa polypeptide. The colour is due to a retinal molecule covalently bound as a Schiff base to a lysine side chain (Fig. 6.13). A Schiff base has a dissociable proton with a pK_a in aqueous medium of approximately 7 but in the dark-adapted state of bacteriorhodopsin this is elevated to over 10. Such elevation must result from the local environment, for instance a nearby negative charge. It is the reversible association and dissociation of a proton to and from the Schiff base which underlies the ability of bacteriorhodopsin to pump protons. Referring back to Fig. 5.10, a light-driven proton pump must integrate a change in conformation, making the proton binding site (in this case the Schiff base) alternately accessible from the two sides of the membrane, with a change in pK_a. Thus the site must bind protons from the proton-donating side of the membrane (i.e. it must have a high pK_a), and release them on the other side of the membrane (i.e. its pK_a must substantially decrease). During the photocycle the pK_a of the Schiff's base falls substantially, probably below 6. It is thought that one proton per photon is translocated across the membrane, and the following description assumes this stoichiometry.

The retinal group lies approximately in the plane of the membrane attached to Lys-216 at the approximate centre of the bilayer (Plate IV and Fig. 6.13). The structure indicates the presence of a channel from the Schiff base to the external phase and a rather narrower channel to the cytoplasm. Acidic groups provide an obvious means for protons to pass through the protein, and two aspartate residues, Asp-85 on the P-side of the retinal and Asp-96 on the N-side, appear to play a particularly essential role in the translocation. Site-directed mutagenesis of either of these residues to asparagine, or selection of phototrophically incompetent cells of a strain of *Halobacterium* which prove to have the same mutations, drastically restrict the activity of the pump while mutation of other aspartate residues has less or even no effect. *Halobacterium halobium* contains a second retinal-containing protein, *halorhodopsin*, which functions as a Cl^- (but not as a proton) pump. This protein has considerable

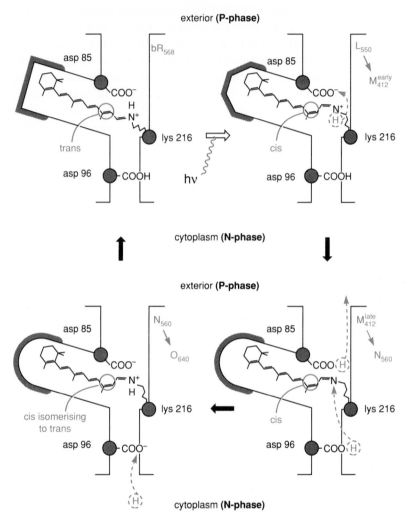

Figure 6.13   Conformational changes of retinal and proton movements in the bacteriorhodopsin photocycle.
Adapted from Henderson *et al.* (1990), omitting other amino acids for simplicity.

sequence similarity with bacteriorhodopsin, but lacks the aspartates equivalent to those at positions 85 and 96, consistent with their role in proton translocation within bacteriorhodopsin.

6.5.1 The photochemical cycle

The initial conformation of the retinal pigment in the dark-adapted form of bacteriorhodopsin known as bR_{568} (Fig. 6.13) has an all-*trans* conformation and can be shown by its resonance Raman spectrum to be protonated on the nitrogen of the Schiff base. Plate IV shows the arrangement of the chromophore within the protein. The carboxyl of Asp-96 in the cytoplasmic channel is believed from Fourier transform infra-red spectroscopy to be protonated, despite the high cytoplasmic pH. This implies that it must be in a special environment capable of increasing its pK_a considerably. There is a precedent for this in enzymes such as lysozyme. In contrast, Asp-85 has an abnormally low pK_a and would be initially deprotonated.

(a) Stage 1: the trans–cis isomerization of the chromophore

When bR_{568} absorbs a photon it forms a transient excited state K in which the excited chromophore undergoes an isomerization from all-*trans* to 13-*cis* (Fig. 6.13 and Table 6.1). This is the only light-dependent reaction, and the other

Table 6.1 Schematic photochemical cycle for bacteriorhodopsin

State	Structure of		pKs of protonatable groups and H⁺ movement				
	Retinal	Protein	Cytoplasm	Asp-96	Schiff	Asp-85	Medium
1. bR_{568}	*trans*	'T'	pH = 7 $[H^+_{\#3}]$	pK = 10 $[H^+_{\#2}]$ ■	pK = 10 $[H^+_{\#1}]$	pK = 3	pH = 7
					$hv\downarrow$		
2. K_{610}	*cis*	'T'	pH = 7 $[H^+_{\#3}]$	pK = 10 $[H^+_{\#2}]$ ■	pK = ? $[H^+_{\#1}]$	pK = 3	pH = 7
3. L_{550}	*cis*	'T'	pH = 7 $[H^+_{\#3}]$	pK = 10 $[H^+_{\#2}]$ ■	pK < 6 $[H^+_{\#1}]$	pK = 3	pH = 7
4. M_{412}^{early}	*cis*	'T'	pH = 7 $[H^+_{\#3}]$	pK = 10 $[H^+_{\#2}]$	pK < 6 ■	pK = 3 → $[H^+_{\#1}]$	pH = 7
5. M_{412}^{late}	*cis*	'C'	pH = 7 $[H^+_{\#3}]$	pK = 10 $[H^+_{\#2}]$	pK = 10 ■	pK = 3 $[H^+_{\#1}]$	pH = 7
6. N_{560}	*cis*	'C'	pH = 7 $[H^+_{\#3}]$	pK = 10 → $[H^+_{\#2}]$	pK = 10 ■	pK = 3 →	pH = 7 $[H^+_{\#1}]$
7. O_{640}	*trans*	'T'	pH = 7 →	pK = 10 $[H^+_{\#3}]$	pK = 10 ■ $[H^+_{\#2}]$	pK = 3	pH = 7 $[H^+_{\#1}]$
8. bR_{568}	*trans*	'T'	pH = 7	pK = 10 $[H^+_{\#3}]$ ■	pK = 10 $[H^+_{\#2}]$	pK = 3	pH = 7 $[H^+_{\#1}]$

The Schiff base and the two Asp residues implicated in proton pumping are shown. The movement of three protons is shown from the cytoplasm at pH 7 to the external medium also at pH 7. (→) proton movements; ■ conformational block. Remember that a group will be protonated if its pK_A is higher than the ambient pH. Thus the high pK_A of Asp-96 is essential if it is to bind protons from a cytoplasm at pH 7. Note that the only change in pK_A is that of the Schiff base itself. The scheme is an oversimplification and other amino acid side chains may participate in protonation/deprotonation events – see, for example, Mathies *et al.* (1991). There is likely to be water in the channels that connect the Schiff base to the two aqueous phases. The pK_a values for the two Asp carboxylates are shown as invariant although they could change during the cycle. The 'T' state is the equilibrated protein conformation with the all-*trans* chromophore; the 'C' state is the equilibrated conformation with a 13-*cis* chromophore. A precursor to K, called J, forms in 500 fs and relaxes to K on a 3-ps timescale. The chromophore structure in J is unclear. The approximate timescale for the other transitions are K → L 2 μs, L → M 60 μs, M → N 2 ms, N/O 2 ms, O → bR 0.5 ms.

stages of the cycle are 'dark reactions' which occur spontaneously, but in some cases at the expense of energy stored in the chromophore, to restore the initial conformation.

The intermediate L now forms (Table 6.1) in which the Schiff base has a greatly decreased affinity for protons, i.e. it possesses a lower pK_a. It has been suggested that this occurs because the Schiff base has been moved into a more hydrophobic environment.

(b) *Deprotonation of the Schiff base*

Proton transfer to Asp-85 is now energetically and kinetically possible and occurs after several microseconds to give a species that absorbs maximally at 412 nm, termed 'M_{412} early' (Table 6.1). It is believed that there then follows a conformational change in the protein to give what is known as the C state (Table 6.1) at the expense of energy stored in the chromophore. This conformational change to give the 'M_{412} late' species also causes the Schiff base to be in communication with the cytoplasm (N-phase) rather than the external aqueous P-phase (Table 6.1). There is clear evidence from time-resolved X-ray diffraction studies that there is a change in the tertiary structure of bacteriorhodospin on going from the dark-adapted state (bR_{568}) to the M_{412} state during the photocycle (Koch *et al.*, 1991). The accessibility of the Schiff base to attack by borohydride or hydroxylamine only under illuminated conditions also indicates conformational change within the polypeptide. Such conformational change would prevent the proton returning from Asp-85 to the Schiff base which is now once again a good proton acceptor with high pK_a (Table 6.1). Asp-85 has a low pK_a and consequently the proton will be released to the P-phase at either this stage or upon the transition to the N state (Table 6.1).

(c) *Reprotonation of the Schiff base*

The transition from M_{412} late to N is associated with the donation of a proton to the Schiff base from Asp 96 (Fig. 6.13 and Table 6.1). The scheme shown in Table 6.1 indicates that the pK_a values of the two aspartate groups do not change. This is not firmly established but there is evidence that the two proton-transfer steps, to and from the Schiff base, occur at ΔG values of close to zero (Varos and Lanyi, 1991), which implies that under the conditions of proton transfer the pairs of pK_a values must be closely matched (cf. Table 6.1). The initial state of the bacteriorhodopsin molecule is then regained by transfer of a proton on to Asp-96 (transition from N to O state) and relaxation of both the chromophore and the polypeptide into the *trans* and T structures. There is a final relaxation step from O to bR_{568} (Table 6.1). Time-resolved X-ray difraction studies support the notion that there is significant conformational change in the protein between the N and bR_{568} states (Koch *et al.*, 1991). Such structural changes are consistent with, but do not prove, the C–T model (Table 6.1).

Measurements of charge translocation across membranes containing bacteriorhodopsin indicate that 80% of the charge movement occurs during the last part of the photocycle as the M state returns to bR with concomitant movement of protons from the cytoplasm to the Schiff base. This is consistent with the proton channel connecting the Schiff base to the cytoplasm being relatively narrow and hydrophobic relative to the equally long but wider channel that connects the Schiff base to the external aqueous phase. On this basis the principal electrical barrier to movement of the proton across the membrane would be presented by the pathway leading from the cytoplasm to the Schiff base.

Finally, it should be emphasized that the proposed structural changes within the photochemical cycle and the pathway for the proton are far from proven. This would require the unambiguous determination of the structure of one or more of the transient intermediates, for example the deprotonated M state. For this approach to succeed the lifetime of this state would have to be prolonged, perhaps through the use of a mutated protein made by site-directed mutagenesis.

Racker assembles the cold-labile F_1–ATP synthase (negatively stained with phosphotungstate), while Mitchell juggles protons and charges, Slater attempts to grasp the elusive squiggle, and Boyer induces a conformational strain

7 THE ATP SYNTHASE

7.1 INTRODUCTION

Reviews Senior 1990, Penefsky and Cross 1991, Boyer 1989, Fillingame 1990

In contrast to the great variety of mechanisms found in different organisms for the respiratory or photosynthetic generation of Δp, the major consumer of Δp, the ATP synthase, is highly conserved. It is present in mitochondria, chloroplasts, both aerobic and photosynthetic bacteria, and even those bacteria which lack a functional respiratory chain and rely on glycolysis. For reasons that will soon become clear, the proton-translocating ATP synthase is known as an $F_1.F_o$-ATPase (or ATP synthase), distinguishing it both from P-type (or E_1-E_2) ATPases such as the (Na^+/K^+)-translocating ATPase in eukaryotic plasma membranes, whose catalytic cycle involves a covalent attachment of the phosphate from ATP, and also from the more recently described class of V (for vacuolar) ATPases which can pump protons across internal membranes (e.g. tonoplasts in plants and synaptic vesicles in neurones).

The function of the ATP synthase is to utilize Δp to maintain the mass-action ratio for the ATPase reaction 7 to 10 orders of magnitude away from equilibrium, or in the case of fermentative bacteria, such as *S. faecalis*, to utilize ATP to maintain Δp for the purpose of transport. Although the function of the complex in all except this last case is to synthesize rather than hydrolyse ATP, it is sometimes referred to as the proton-translocating ATPase.

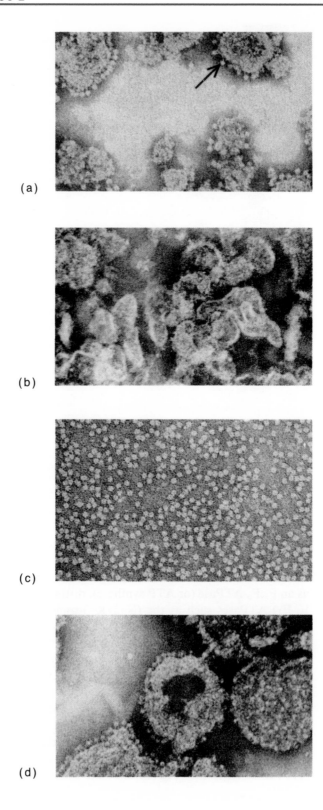

(a)

(b)

(c)

(d)

7.2 F_1 AND F_o

The general features of the ATP synthase were deduced from early studies with submitochondrial particles (SMPs) (Section 1.3.1). These inside-out membrane vesicles catalyse either ATP synthesis in the presence of a respiratory substrate or ATP hydrolysis in the absence of such a substrate. Both respiration and ATP hydrolysis generate a Δp (with the positive, P-phase, in the lumen of the vesicle) and H^+/ATP ratios can be determined, albeit imprecisely. In good preparations of SMPs the rate of ATP hydrolysis can be stimulated by addition of a protonophore to collapse Δp.

Both the synthesis and hydrolysis of ATP by mitochondrial membranes can be inhibited by several reagents, including oligomycin and dicyclohexylcarbodiimide (DCCD); at characteristically low concentrations the latter reacts with a particular glutamate residue in a hydrophobic environment.

The ATP synthase can be visualized under the electron microscope in preparations of SMPs which have been negatively stained with phosphotungstate. The complexes appear as roughly spherical knobs projecting from the original matrix side of the membrane (Fig. 7.1). When SMPs were washed with urea, chelating agents or low ionic strength media the knobs were lost from the membrane (Fig. 7.1). At the same time ATPase activity was solubilized from the SMPs. This activity was inhibited neither by oligomycin nor DCCD and was termed the F_1-ATPase ('fraction 1'). Naturally the soluble F_1 was incapable of ATP synthesis. The stripped SMP membranes had lost ATP synthase and ATPase activity but interestingly also behaved in an 'uncoupled' manner, with no respiratory control with NADH as substrate and with evidence of a high proton permeability. When these depleted SMPs were pretreated with either DCCD or oligomycin some respiratory control was reintroduced, and the proton permeability, estimated, for example, from the rate of decay of a pH gradient after cessation of respiration, was reduced to the considerably lower value seen with untreated SMPs.

These seminal observations suggested that oligomycin and DCCD bind to and inhibit a component in the membrane of depleted SMPs which could conduct protons and that in intact SMPs the free passage of protons through this component was in some way controlled by F_1. This proton channel was termed

Figure 7.1   The F_1 component of the mitochondrial ATP synthase visualized by negative staining of submitochondrial particle membranes.
(a) Electron micrograph of submitochondrial particles (magnification $175\,000\times$) shows F_1 molecules (arrow) on the surfaces. (b) Electron micrograph showing particles from which F_1 has been removed by urea treatment; the particles have greatly diminished ATPase activity. (c) Molecules of F_1 visualized by electron microscopy after removal from submitochondrial particles. (d) The appearance of urea-treated particles following reconstruction with F_1; the surface particles are again evident and ATPase activity is restored. (Reproduced with permission from G. Weissmann and R. Claiborne (eds) *Cell Membranes: Biochemistry, Cell Biology and Pathology* (1975)).

F_o (fraction oligomycin) and can be inhibited by oligomycin and DCCD in both depleted and untreated SMPs. Clearly the soluble F_1-ATPase will not be sensitive to these inhibitors. Purification of the intact $F_1.F_o$-ATP synthase (ATPase) complex requires detergents to maintain the solubility of the highly hydrophobic F_o.

These observations can be generalized to bacterial and thylakoid membranes, although whereas DCCD is generally effective in all systems, oligomycin, as well as a third inhibitor venturicidin, only inhibits the $F_1.F_o$-ATP synthase (ATPase) from mitochondria and a limited number of bacterial genera. In all energy-conserving membranes ATP is always hydrolysed or synthesized on the side of the membrane from which the knobs project (the N-phase), while during ATP synthesis protons cross from the side which lacks knobs (the P-phase, Fig. 1.1). Thus F_1 faces the mitochondrial matrix, the bacterial cytoplasm and the chloroplast stroma.

7.3 THE SUBUNITS OF THE $F_1.F_o$-ATP SYNTHASE (OR ATPase)

The most definitive information available is for the *E. coli* $F_o.F_1$ ATP synthase. F_1 preparations have five types of polypeptide, usually known as α, β, γ, δ and ε, whilst three further polypeptide species, a, b and c, have been identified in F_o. The gene for the *E.coli* $F_o.F_1$ATP synthase is a single operon containing eight cistrons that correspond to the eight polypeptides. The stoichiometry of the subunits of the *E. coli* $F_1.F_o$-ATP synthase is $ab_2c_{10-12}\alpha_3\beta_3\gamma\delta\varepsilon$. The estimated molecular masses of F_o and F_1 are estimated to be 160 kDa and 370 kDa respectively. Amino acid sequences have been deduced from DNA sequences for all eight polypeptides. The five F_1 polypeptides are consistent with a hydrophilic globular structure, while each of the F_o polypeptides has hydrophobic regions consistent with transmembrane α-helices.

ATP synthases from other sources have very similar subunit structures, although those from mitochondria and thylakoids are more complex, possessing extra subunits, the functions of which are unclear. There is not an exact correspondence between the roles of the smaller subunits in the enzymes from different sources. Thus the *E. coli* δ-subunit is equivalent to the mitochondrial oligomycin-sensitivity-conferring factor (OSCP), a subunit of F_1 that is required for the mitochondrial $F_1.F_o$ to be sensitive to oligomycin. The mitochondrial δ-subunit is related to the ε-subunit of the *E. coli* enzyme.

7.4 THE STRUCTURE OF F_o

Figure 7.2 shows a speculative model for the structure of the $F_1.F_o$-ATPase, the evidence for which will now be reviewed, starting with the hydrophobic F_o. The justification for the extended stalk region is not just the early electron

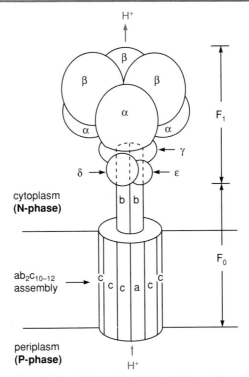

Figure 7.2   A structural model of the $F_1.F_o$ ATP synthase from _E. coli_.
The F_1 sector has an estimated diameter of approximately 90 Å. The stalk,
which may be as long as 45 Å (Gogol _et al._ 1987) is shown here as mainly
comprising the _b_-subunits. The α and β subunits are arranged in an alternating hexagonal array but are possibly more elongated than shown here.
This would accommodate experimental findings such as a reported cross-
linking of the β subunit to F_1 to the a-subunit of F_o (see Penefsky and
Cross, 1991, for further details).

microscopic observations (Fig. 7.1) but also more recent cryo-electron microscopy of unstained molecules which clearly indicates that F_1 is connected to F_o
by an extended narrow spindle of protein (Gogol _et al._, 1987).

   The sequence of the a-subunit of F_o suggests six to eight transmembrane
α-helices but little can be said with certainty about the structure or function
of this subunit. Although the b-subunit of F_o has a hydrophobic N-terminus
the rest of the sequence is hydrophilic: it has been suggested that the latter
could form a hydrophilic spindle onto which the F_1 complex is mounted,
consistent with the structural information from cryo-electron microscopy.

   The c-subunit of _E. coli_ has an aspartate residue (_E. coli_ is atypical in this
respect, glutamate being found in other sources) in an otherwise hydrophobic
sequence that is predicted to lie in the middle of the bilayer: this is the residue
to which DCCD binds. _E.coli_ mutants in which the aspartate is changed to
asparagine cannot conduct protons through F_o. It is clearly tempting to propose an essential role for this residue in proton translocation, although this
cannot be regarded as proved. A comparable role for acidic amino acid
residues has been proposed for bacteriorhodopsin (Section 6.5).

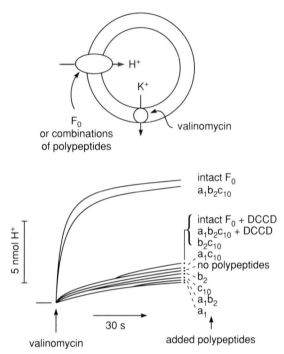

Figure 7.3 All three F_o subunits from _E. coli_ are required to create a proton channel in a phospholipid bilayer.
E. coli F_o or various components of it in the ratios shown were incoporated into phospholipid vesicles preloaded with K^+. The reaction was started by adding valinomycin and the proton influx monitored with a pH electrode. The inhibitor DCCD, which binds to the c subunit of F_o, was present as shown. (From Schneider and Altendorf, 1987).

All three F_o subunits are required in approximately the _in situ_ ab_2c_{10-12} stoichiometry to create a proton channel in a phospholipid bilayer (Fig. 7.3). Note that valinomycin is required in this experiment to allow an efflux of K^+ to compensate for the charge on the protons entering through the channel. Using an F_o preparation from the thermophilic bacterium PS3 it has been shown that ion flux decreases with increasing pH, suggesting, but not proving, that protons rather than hydroxyl ions are the transported species.

7.5 THE STRUCTURE OF F_1

A variety of experimental evidence, including X-ray diffraction data at 0.36 nm (Bianchet _et al._, 1991), indicates that the α- and β-subunits are arranged alternately in an assembly that has three-fold symmetry, consistent with the $\alpha_3\beta_3\gamma\delta\varepsilon$ stoichiometry. The β-chain from many different sources (e.g. mitochondria, plants and bacteria) contains highly conserved sequences characteristic of adenine nucleotide binding sites in other proteins. Chemical modification of the β-subunit, e.g. by Nbf-Cl or photoaffinity labels, inactivates the enzyme, while the purified β-subunits from some sources may have ATPase activity in the absence of other subunits. Such indications that the β-subunit is the site of ATP synthesis will only be rigorously proven by a high-resolution structural determination. It must be also borne in mind that adenine nucleotide binding sites may be located at an interface between two subunits, particularly since the α- and β-subunits appear to alternate in the structure (Fig. 7.2).

Whilst conservation of sequence is greatest amongst β-subunits it is also pronounced amongst α-chains, which also have sequences indicative of adenine nucleotide binding sites. Sequence conservation is less pronounced amongst the remaining subunits. The role of the γ-, δ- and ε-subunits is unclear, except that they are all needed for the reconstitution of an $F_1.F_o$-ATP competent in proton translocation. It is widely assumed that they contribute to the stalk region seen in electron microscopy, whilst in the case of the thylakoid enzyme there is evidence that the γ-subunit controls proton flow through the enzyme.

7.6 THE MECHANISM OF ATP SYNTHESIS

The mechanism by which proton translocation is coupled to ATP synthesis by $F_1.F_o$-ATP synthase is not known in molecular detail, although there is considerable information on the chemistry of the covalent bond formation between ADP and Pi. The mechanism is fundamentally different from that of the P-type-ATPases such as the (Na^+/K^+)-ATPase of mammalian plasma membranes where the enzyme becomes transiently phosphorylated; no phosphorylated intermediate is detected in the ATP synthase. It should be recalled that the stoichiometry of proton translocation H^+/ATP by the ATP synthase is not known for certain (Section 4.5) and such uncertainty does not help in the quest for the mechanism. The balance of experimental evidence suggests that the ratio is 3 and further support for this value has come from experiments in which the complete $F_1.F_o$ ATP synthase from a thermophilic bacterium has been incorporated into a planar phospholipid bilayer (Section 1.3.5) (Hirata *et al.*, 1986). An ATP-dependent current could be detected but an applied transmembrane potential of 180 mV, negative in the N-phase compartment (into which F_1 projects), prevented the current. The thermodynamics of the ATP hydrolysis reaction under these conditions was consistent with an H^+/ATP ratio of at least 3.

7.6.1 Overview

To provide a framework for the subsequent discussion we shall start by describing a hypothetical model, which is no more than a working hypothesis, and then proceed to discuss the evidence for such a model. We shall always refer to the compartment into which F_1 projects as the N-phase and the compartment from which the protons enter F_o during ATP synthesis as the P-phase.

When protons cross the membrane from the P-phase through F_o they pass down the electrical field created by the membrane potential and accumulate at the bottom of the F_o proton 'well' (Fig. 7.4). Thus regardless of whether the major component of Δp is a $\Delta \psi$ as in the case of mitochondria, or a ΔpH as for the thylakoid, a large proton concentration gradient will be established between the bottom of F_o and the N-phase. How might this concentration gradient be exploited to generate a conformational change in F_1 which can subsequently drive ATP synthesis? As in the case of the hypothetical redox-driven proton pump discussed in Chapter 5 and bacteriorhodopsin discussed in Chapter 6, it is necessary to co-ordinate changes in proton binding affinity with changes in conformation which both drive the synthesis of ATP and make the proton binding site(s) alternately accessible to the P-phase and the N-phase.

Since the protons are at high concentration at the bottom of the F_o well, they would bind to an accessible site on F_1 even if it has a relatively low pK_a. Conversely the proton will be able to dissociate from the binding site following a hypothetical conformational change which makes the site accessible to the N-phase even if in the process the proton becomes much more tightly bound (i.e. the pK_a of the site increases), since the pH of the N-phase may be 3–4 pH units higher than at the bottom of the F_o well. This change from a loosely to a tightly bound proton means that the proton loses Gibbs energy which can be transferred to the protein, allowing it to adopt a higher energy conformation.

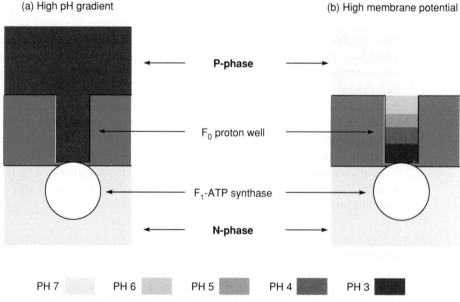

(a) High pH gradient   (b) High membrane potential

P-phase

F_o proton well

F_1-ATP synthase

N-phase

PH 7   PH 6   PH 5   PH 4   PH 3

Figure 7.4   The F_o channel functions as a proton well transducing a $\Delta \psi$ to a ΔpH.

Thus one can readily devise a model whereby a protonmotive force can induce a 'high-energy state' in a protein: how can this be used to drive ATP synthesis? As we shall see, the most plausible models suggest that the proton-induced conformational changes are integrated with changes in binding affinity of the adenine nucleotide binding such that very tightly bound ADP and Pi form ATP with little input of Gibbs energy, but that the Δp-induced 'high energy state' of the complex is required to lower the affinity of the binding site and allow the ATP to leave the complex. Finally the initial conformation must be regained. Thus there is an effective alternate exposure of the proton-binding site to the two sides of the membrane, as occurs for the Schiff's base of the retinal in bacteriorhodopsin (Section 6.5). We shall now discuss the evidence for this model.

7.6.2 ATP hydrolysis by soluble F_1

(a) Mechanism

ATP hydrolysis is more easily studied than ATP synthesis, since the former only requires soluble F_1, although of course no information on proton coupling can be obtained in this membrane-free preparation.

The hydrolysis of ATP involves breakage of the bond between the bridging oxygen and the γ-phosphorus atom and thus an oxygen on the liberated Pi comes from water (Fig. 7.5). This can be shown by carrying out the hydrolysis in $H_2{}^{18}O$ and quantifying the label appearing in Pi. Under some conditions the hydrolysis of ATP by SMPs results in more ^{18}O incorporation into Pi than can be accounted for by ATP hydrolysis alone. One explanation for this is that Pi, produced at the active site with one ^{18}O by hydrolysis of ATP, can rotate within the active site before being released. If the ATP hydrolysis reaction is reversible, and if the rate of release of products from the catalytic site is slow relative to the rate of resynthesis of ATP, then the Pi reincorporated into ATP will lose ^{16}O rather than ^{18}O. Subsequent ATP hydrolysis will lead to further incorporation of ^{18}O into Pi which will ultimately be released to the medium. Most surprisingly, this extra incorporation of ^{18}O was not abolished when Δp was dissipated by a protonophore.

The occurrence of this exchange with SMPs in the absence of a Δp was one of the first pieces of evidence that the protonmotive force was not directly used to condense ADP and Pi. In subsequent experiments the analogous exchange reaction was studied with the F_1 fragment but with the difference that water was unlabelled the substrate ATP was labelled on its γ oxygens with ^{18}O and the deficit of ^{18}O in the product phosphate was analysed. This procedure was more sensitive. When the hydrolysis of extremely low concentrations of ATP (less than 10^{-6} M) was studied, the loss of ^{18}O label from product Pi was considerably more than expected from the hydrolysis reaction alone, again indicating that exchange had occurred within a catalytic site. This provided further evidence that the hydrolysis of ATP to ADP and Pi by the soluble F_1 was to

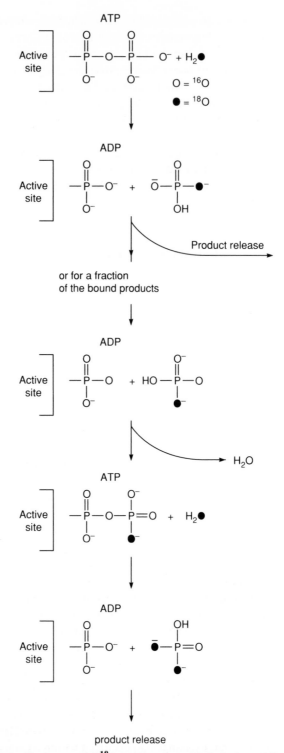

Figure 7.5 Exchange of O^{18} into Pi or from H_2O during ATP hydrolysis by F_1-ATPase indicates that the reaction is reversible.

some extent reversible even with no input of energy from Δp (Fig. 7.5). This conclusion could be reinforced by similar experiments in which loss by exchange of ^{18}O originally in Pi to water was observed during ATP synthesis in steady-state oxidative phosphorylation. More complex experiments (see Boyer, 1989) in which these exchange reactions were studied with variable substrate concentrations indicate that, in addition, interactions occur between more than one catalytic site in the ATP synthase.

An equilibrium constant is equal to the ratio of the forward and reverse rate constants. Since the K' for ATP hydrolysis in free solution is approximately 10^5 M (Section 3.2.1) the reverse reaction (k_{-1} in Fig. 7.5) should be undetectably slow. F_1 therefore appears to alter the equilibrium constant for ATP hydrolysis to make the rate of the reverse reaction significant. How can this occur without offending against the first law of thermodynamics, since we appear to be making ATP with no energy input? The answer is that we are not making *free* ATP, but rather ATP bound to F_1. If that ATP is very firmly bound then much energy will have to be used to release the ATP into solution. Consider the ΔG values associated with the following steps:

1. $$ADP_{free} \rightleftharpoons ADP_{bound} \qquad \Delta G_1$$

2. $$Pi_{free} \rightleftharpoons Pi_{bound} \qquad \Delta G_1$$

3. $$ADP_{bound} + Pi_{bound} \rightleftharpoons ATP_{bound} \qquad \Delta G_3$$

4. $$ATP_{bound} \rightleftharpoons ATP_{free} \qquad \Delta G_4$$

The overall reaction is the sum of these steps:

$$ADP_{free} + Pi_{free} \rightleftharpoons ATP_{free} \qquad \Sigma\Delta G = +40 \text{ kJ mol}^{-1*}$$

*typical value for the mitochondrial matrix

What is being observed in the ^{18}O exchange reaction is not the overall reaction, but reaction 3, which occurs with a ΔG close to zero. Nearly all the input of $+40$ kJ mol^{-1} is required for the final step, the removal of very tightly bound ATP from the catalytic site. As we shall see below, it is the conformational change driven by the protonmotive force (see above) which releases the bound ATP. Confirmation of this model has come from measurements of the dissociation constant for the F_1.enzyme.ATP complex, where a value of about 10^{-12} M was obtained.

With very low concentrations of ATP ($<10^{-10}$ M) labelled with ^{32}P on the γ-phosphate, hydrolysis catalysed by F_1 proceeds very slowly and an equilibrium mixture of ATP, ADP and Pi will be at the catalytic site. If, however, a higher concentration of cold ATP is subsequently added in a chase experiment the rate of hydrolysis of the already bound ATP-$\gamma^{32}P$ considerably increases. It appears that the higher concentration of ATP occupies one or more lower affinity ATP binding sites on different β-subunits of the enzyme and that this causes a conformational change, allowing hydrolysis of ATP and release of the products from the high affinity site at an accelerated rate. Thus there are site-to-site interactions, perhaps mediated through subunit interfaces. This negative co-operativity of binding but positive co-operativity of catalysis (k_{cat}) explains why the ^{18}O exchange experiment discussed above must be performed at very low ATP concentrations: at higher concentrations of ATP the ADP formed would be released rather than remaining to allow the reverse reaction (and thus ^{18}O exchanges) to occur. It is notable that mutations in the α-chain of the *E. coli* enzyme attenuate such co-operativity and impair $\alpha\beta$ interface interactions.

The stereochemistry of ATP hydrolysis can be investigated by isotopic labelling. We know from the ^{18}O exchange experiment that it is the bond between the γ-P and oxygen which is broken because ^{18}O from $H_2{}^{18}O$ is incorporated into Pi rather than ADP. The attack of water on the γ-phosphate could be similar to a classic SN2 reaction at a saturated carbon atom with the phosphate inverting its configuration (Fig. 7.6). This cannot be tested with ATP itself since the γ-phosphate is not a chiral centre; however, the ATP analogue, Fig. 7.6 , does have such a centre and is a substrate for F_1. Using this analogue together with $H_2{}^{17}O$ confirms this inversion. In contrast to F_1, which does not form a phosphoenzyme intermediate, a P-type-ATPase such as the (Na^+/K^+)-ATPase which does form such an intermediate retains the stereochemistry of the terminal phosphate. In the latter enzymes two inversions corresponding to phosphorylation and dephosphorylation steps occur, leading to net retention of configuration.

(b) *Structure–function interrelationships*

F_1 is thought to have a total of six adenine nucleotide binding sites. Only for

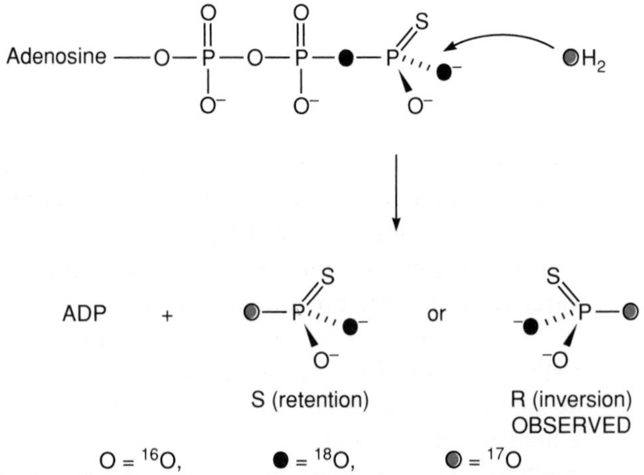

Figure 7.6 Stereochemistry of ATP hydrolysis by F_1 ATPase.
As shown, adenosine 5'-(3-thiotriphosphate), stereospecifically labelled with ^{18}O in the γ position, was hydrolysed in $H_2{}^{17}O$ to give a chiral $^{16,17,18}O$ Pi product. The configuration of the product showed that hydrolysis had proceeded with inversion at the γ phosphorus atom.

three (or possibly two) of these is there any evidence for catalytic function as discussed above. The role of the other sites, which may be either on α-chains or the $\alpha\beta$ interfaces, and from which exchange is very slow, continues to be uncertain.

With the exception of DCCD, which has identified an essential aspartate residue in the c-subunit of F_o, chemical modification has provided limited information on the mechanism of the ATP synthase. Thus, although Nbf-Cl inactivates F_1 by reacting with a conserved Tyr residue (311 in the beef heart enzyme) on a single β-subunit changing this residue to Phe by site-directed mutagenesis in yeast or *E. coli* does not significantly alter the enzyme activity. Thus the hydroxyl of Tyr is not essential, and the reason for the inhibitory action of Nbf-Cl may be the steric hindrance caused by the bulky modifying group. The nitrobenzofurazan group from Nbf-Cl can undergo an intra-molecular shift within a β-chain from its initial site at Tyr_{311} to Lys_{162}, indicating that the residues must be adjacent. The enzyme remains inactive. This is notable since Lys_{162} is found at a sequence motif -Gly-X-X-X-X-Gly-Lys- which is found in many ATP binding enzymes and is thought to participate in binding of phosphate groups. In accord with this possibility Pi will protect F_1 against the initial labelling by Nbf-Cl.

The implication of Lys-162 at an active site is strengthened by the high-resolution X-ray structure of a ras protein known as p21. This protein catalyses the slow hydrolysis of GTP, and the amino group of the Lys in the -Gly-X-X-X-X-Gly-Lys- motif (this is commonly called the P loop) is co-ordinated to an oxygen atom for each of the β- and γ-phosphate groups (Wittinghofer and Pai, 1991). Hydrolysis of GTP occurs through an associative in-line nucleophilic attack of water and the crystal structure shows that there is a water molecule positioned for such attack. The lysine is attached to an oxygen of the β-phosphate of GDP after the hydrolysis. It is probable that the hydrolysis, and in reverse the synthesis, of ATP will occur via a similar mechanism in F_1. However, such conclusions must be tempered by some uncertainties generated by site-directed mutagenesis experiments. Thus while the equivalent Lys in the *E. coli* enzyme can be changed to the oppositely charged Glu with little loss in activity (contrary to expectation if the lysine's polarizing power is important), F_1 from the bacterium PS3 is inactivated when the Lys is changed to isoleucine.

The $\alpha_3\beta_3\gamma\delta\varepsilon$ stoichiometry of F_1 would suggest that the complex has three catalytic sites corresponding to the three β-subunits. However, each of the α- and β-subunits must make different contacts with the single-copy γ-, δ- and ε-subunits since these latter show no evidence of any tripartite repeating sequences. An asymmetry is also apparent because Nbf-Cl inactivates by binding to a single β-chain and because F_1 shows high- and low-affinity ATP binding sites on different β-subunits (see above). These asymmetric properties persist in an F_1 from bacterium PS3 from which the single-copy subunits have been removed to give a $\alpha_3\beta_3$ structure. Hence it is not clear either whether the binding of nucleotides or Nbf induces the asymmetry or whether the $\alpha_3\beta_3$ structure is intrinsically asymmetric.

7.6.3 ATP synthesis by membrane-bound $F_1.F_o$ ATP synthase

The study of ATP hydrolysis by soluble F_1 is of value to the extent that it provides information about ATP synthesis by the membrane-bound $F_1.F_o$ ATP synthase. The ^{18}O exchanges discussed above can also be observed with the intact $F_1.F_o$ complex *in situ* in SMPs in the absence of Δp. However, for reasons which are not clear it is no longer necessary to use a very low ATP concentration in order to see the exchange.

With the intact complex, blockade of the F_o proton channel by either oligomycin or DCCD drastically reduces the high-affinity binding of ATP to F_1, indicating a conformational change relayed to an active site through the subunits of the enzyme from the site of inhibitor binding.

If the molecular basis of ATP synthesis is the reverse of the F_1 hydrolysis discussed above, it follows that the major ΔG changes are associated with the binding of ADP and Pi and/or the release of ATP, and that these are the steps which must in some way be coupled to Δp through a conformational change. A crucial experiment using SMPs involved loading the very high affinity ATP binding site on F_1 in the absence of a Δp and then initiating respiration. The generated Δp caused a release of this tightly bound ATP, clearly demonstrating that Δp decreases the binding affinity of ATP. The change in binding affinity has to be dramatic: from a value of about 10^{-12} M in the absence of Δp to a sufficiently loose binding such that ATP can dissociate in the presence of the normal N-phase concentration of the nucleotide. Experiments with a bacterial vesicle system have suggested that during ATP synthesis the competitive binding of ATP to the catalytic site has a dissociation constant of 10^{-5} M (Perez and Ferguson, 1990), a change of 10^7 induced by the proton-motive force.

Although we do not know how proton translocation is coupled to these affinity changes it is possible to create a plausible model (Fig. 7.7). It is assumed (but not certain) that the H^+/ATP stoichiometry is 3 and that the proton binding sites can exist in a low-affinity state facing the P-phase and a high-affinity state facing the N-phase. It is also assumed that there are three catalytic sites for nucleotide which can exist in O (open), L (loose) and T (tight) conformations and that the proton-induced conformational change causes a T-site with bound ATP to become an O-site and release its bound ATP and at the same time causes a second site to change from an L-site, with loosely bound ADP and Pi, to a T-site where the substrates are tightly bound, allowing bound ATP to be formed. Thus each of these catalytic sites has at any instant a distinct conformation but all the sites pass sequentially through the same conformations. It should be noted that unlike bacteriorhodopsin (Section 6.5), whose reaction mechanism involves proton flux through the catalytic site (i.e. the Schiff's base of the retinal), the balance of evidence is against a direct involvement of the translocated protons in the ATP synthesis reaction itself.

As discussed in Section 5.13.5 ATP synthesis in the bacterium *P. modestum* can be driven by a Na^+ electrochemical potential. The Na^+-translocating ATP

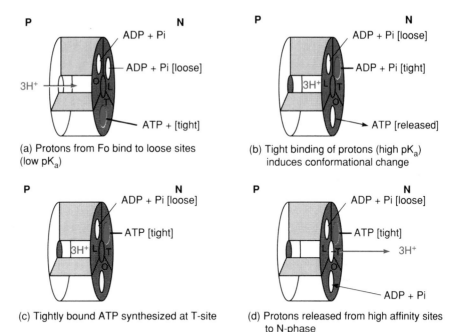

(a) Protons from Fo bind to loose sites (low pK_a)

(b) Tight binding of protons (high pK_a) induces conformational change

(c) Tightly bound ATP synthesized at T-site

(d) Protons released from high affinity sites to N-phase

Figure 7.7   A model for the coupling of Δp to the synthesis of ATP by the F$_1$.F$_0$ ATP synthase.
As described in the text, transfer of protons from low affinity sites (low pK_a) to high affinity sites (high pK_a) is shown as inducing conformational changes in catalytic subunits (here assumed to be three in number and on β chains. O (open), L (loose) and T (tight) indicate three different conformational states of the β subunits. Each subunit experiences each conformation sequentially.

synthase is very closely related to the H$^+$-translocating enzyme, to the extent that F$_1$ subunits are interchangeable with those from *E. coli*. Furthermore DCCD inhibits by binding to the c-subunit. The F$_0$ subunits alone of the *P. modestum* enzyme confer the Na$^+$ translocation property on the F$_1$.F$_0$ enzyme. Amino acid side chains have a relatively weak affinity for Na$^+$ compared with H$^+$. However, the concentrations of Na$^+$ encountered are typically millimolar in contrast to the submicromolar protons, and thus similar changes in cation binding affinity may occur in the *P. modestum* ATP synthase catalytic cycle. The ability of an F$_0$ channel with little apparent modification to transport Na$^+$ instead of H$^+$ could indicate that it is a water-filled channel through which the ions move.

In concluding this section it must be emphasized that knowledge of how the ATP synthase makes ATP is limited by the lack of structural information. A high-resolution structure of the complete F$_1$.F$_0$ structure would not necessarily reveal how the enzyme works. But a structure might prove incompatible with some present ideas. It might, for instance help decide the number of catalytic sites. Although the presence of three β-chains is suggestive of three catalytic sites, this stoichiometry is not unambiguously established. There is no useful model system; a proton-translocating pyrophosphatase found in the cytoplasmic membrane of *Rhodospirillum rubrum*, and which might in principle be simpler than the ATP synthase, has also proved refractory to molecular studies although progress is beginning to be made (Baltscheffsky and Baltscheffsky, 1992).

7.7 NON-THERMODYNAMIC REGULATION OF THE ATP SYNTHASE

The ATP synthases from various sources are regulated in several ways. The mitochondrial enzyme has an additional subunit known as the inhibitor protein. The binding of one molecule to a β-subunit is sufficient to block ATP hydrolysis in the absence of a Δp (Harris and Das, 1991). The inhibition is removed in the presence of a Δp, and thus the inhibitor protein may act to prevent unwanted ATP hydrolysis by mitochondria under conditions such as anoxia when the ATP synthase would otherwise reverse. The regulation in bacteria is less well understood. In the case of *P. denitrificans* and many other organisms in which respiration and hence oxidative phosphorylation is obligatory, ATP hydrolysis and an ATP–Pi exchange reaction is very feeble in the absence or presence of a Δp. It is not known if these bacteria possess inhibitor proteins.

In thylakoids an essential requirement is to avoid ATP hydrolysis in the dark. In the light, activation of the ATP synthase follows exposure of a disulphide bridge in the γ-subunit and its reduction by a thioredoxin which in turn is reduced by ferredoxin. The latter is reduced by the activity of photosystem I (Section 6.4.3). In the dark, ATP synthase relaxes back into an inactive state because reduced ferredoxin and thioredoxin are no longer formed and the disulphide bridge reforms under the more oxidizing conditions. A dimer of a mercaptohistidine compound may be the direct oxidant for the cysteine thiol groups that reform the disulphide bridge (Selman-Reimer *et al.*, 1991).

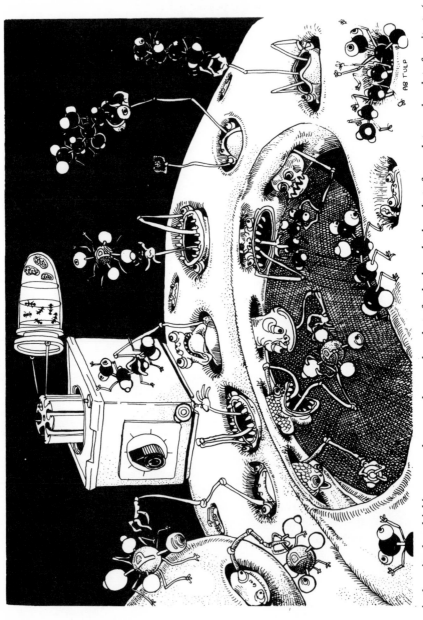

Distinct carriers in the mitochondrial inner membrane exchange phosphate for hydroxyl, phosphate for malate, and malate for citrate (plus a proton)

8 SECONDARY TRANSPORT

8.1 INTRODUCTION

Mitochondria and chloroplasts require a continual interchange of metabolites and end-products with the cell cytosol, whilst in the case of bacteria the interchange is with the external environment. At the same time the membranes must maintain a high Δp for ATP synthesis. Consequently, transport mechanisms are not only designed to operate under the constraints of a high $\Delta\psi$ and/or ΔpH gradient, but also more often than not exploit these gradients to drive the accumulation of substrates or expulsion of products across the membrane. Some of the more common strategies (Fig. 8.1) are:

(a) Proton symport with a neutral species leading to an accumulation driven by the full Δp, e.g. the *lac* transporter of *E. coli*.

(b) Electroneutral proton symport or hydroxyl antiport leading to an accumulation driven by ΔpH alone, e.g. the mitochondrial Pi^-/OH^- exchanger.

(c) Electrical uniport of a cation, driven by $\Delta\psi$, e.g. mitochondrial Ca^{2+} accumulation.

(d) Electroneutral or electrogenic exchange of two metabolites, e.g. the mitochondrial ATP^{4-}/ADP^{3-} antiporter. This is a common device to allow the entry of a polyanionic species such as $citrate^{3-}$ which would be effectively excluded from the mitochondrial matrix due to the high negative membrane potential if it attempted to enter the mitochondrion by a uniport mechanism.

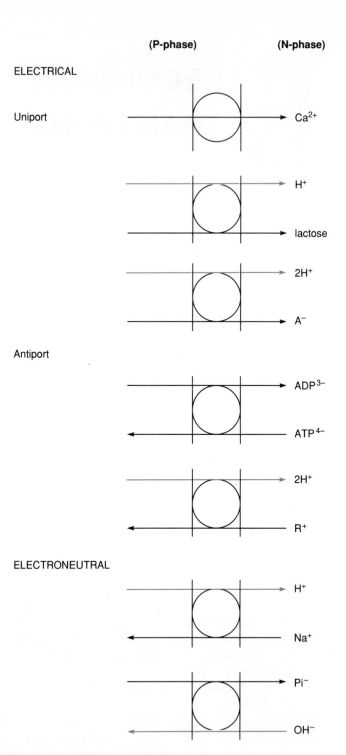

Figure 8.1 Strategies for ion and metabolite transport across energy-transducing membranes.
A^- is a monovalent anionic species: R^+ is a monovalent cation (e.g. a protonated catecholamine, see Endpiece, Chapter 9).

(e) Electroneutral antiport of an ion or metabolite with protons, e.g. the mitochondrial K^+/H^+ antiporter, which expels any excess K^+ which leaks into the matrix in response to the membrane potential.

Mitochondria and bacteria have very distinctive transport properties and these will be considered separately.

8.2 MITOCHONDRIAL MONOVALENT CATION CARRIERS

Selected references   Garlid 1988, Jezek *et al.* 1990

As discussed in Chapter 3 the presence of a 180-mV membrane potential, negative within the matrix, could lead to a 1000-fold accumulation of a monovalent cation which is permeable by uniport. Particularly in the case of K^+, which is present in the cytoplasm at > 100 mM, the slightest uncompensated leakage pathway for the cation into the matrix would lead to catastrophic swelling of the matrix. For this reason mitochondria possess a transporter capable of exchanging either K^+ or Na^+ for H^+. The antiporter has been partially purified and reconstituted. The exchanger automatically expels either cation from the matrix. Since mitochondria operate with a ΔpH of about -0.5 (Section 4.2.4), equivalent to a three-fold gradient of H^+ out/in, this exchanger is capable of lowering the matrix K^+ to about $\frac{1}{3}$ of the concentration in the cytoplasm.

8.3 MITOCHONDRIAL Ca^{2+} TRANSPORT

Reviews   Nicholls and Ackerman 1982, McCormack *et al.* 1990

Mitochondria from vertebrate sources possess potent pathways for the uptake and efflux of Ca^{2+}. When exposed to Ca^{2+} concentrations in excess of 1 μM mitochondria accumulate the cation into their matrix, via a uniport mechanism. The Ca^{2+} uniporter, which remains to be characterized at the molecular level, transports the divalent cation, and is thus in theory capable of supporting a 10-fold gradient of the free cation for each 30 mV of $\Delta\psi$. A typical potential of 180 mV could therefore result in an equilibrium concentration gradient of no less than 10^6. This does not occur in practice, however, since the mitochondrial inner membrane also possesses an independent efflux pathway, functioning as a $Ca^{2+}/2Na^+$ antiporter in mitochondria from most tissues, including heart, brain and brown adipose tissue, and as a $Ca^{2+}/2H^+$ antiporter in liver. Either mechanism drives the efflux of Ca^{2+} from the matrix, the Na^+-coupled pathway requiring the additional intervention of the Na^+/H^+ exchanger introduced above (Fig. 8.2).

The existence of the independent efflux pathway can be demonstrated most simply by the selective inhibition of the uptake pathway once steady-state conditions have been obtained. One such inhibitor is the glycoprotein stain Ruthenium Red, which inhibits the uniporter at very low concentrations. As

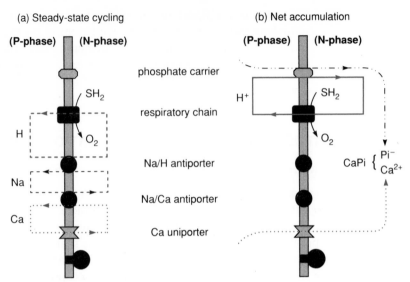

Figure 8.2 Ca^{2+} cycling and accumulation across the inner membrane of heart mitochondria.
(a) Steady-state cycling of Ca^{2+}: a circuit of protons (via respiratory chain and Na$^+$/H$^+$ antiporter) drives a Na$^+$ circuit (via the Na$^+$/H$^+$ antiporter and the 2Na$^+$/Ca^{2+} antiporter) which in turn drives a Ca^{2+} circuit (via the 2Na$^+$/Ca^{2+} and the Ca^{2+} uniporter. A steady-state Ca^{2+} distribution is attained when the kinetics of the Ca^{2+} uniporter and 2Na$^+$/Ca^{2+} antiporter are equal and opposite. This in turn is largely determined by the cytoplasmic free Ca^{2+} concentration, [Ca^{2+}]$_c$, and the value at which a steady state is attained is called the 'set-point'. (b) Net accumulation of Ca^{2+} occurs when [Ca^{2+}]$_c$ increases above the set-point. Massive amounts of Ca^{2+} can be accumulated in parallel with Pi; the two form an osmotically inactive but rapidly dissociable Ca^{2+}–Pi gel.

the presence of this inhibitor does not affect the efflux pathway, a net efflux of Ca^{2+} from the matrix occurs; $\Delta\psi$ would be too high for efflux to occur by reversal of the uniporter even in the absence of Ruthenium Red. The uniporter may also be inhibited by other polycations, such as Mg^{2+} and lanthanides (which also affect the efflux pathway at higher concentrations).

Ultimately what the two carriers set up is an apparently symmetrical cycling of Ca^{2+} across the inner membrane driven by the proton circuit (Fig. 8.2). While this might be thought of as an energy-dissipating, 'uncoupling', process, in practice the rate of Ca^{2+} cycling is restricted by the relatively low activities of the influx and efflux pathways under normal physiological conditions, such that no more than a few per cent of the resting state 4 respiration is used to maintain Ca^{2+} cycling.

8.3.1 Mitochondrial Ca^{2+} transport and the protection of the cell against Ca^{2+} overload

The ability of mitochondria to accumulate Ca^{2+} from media containing

$>1 \mu M$ Ca^{2+} is truly spectacular: in excess of $1 \mu mol$ of Ca^{2+} per mg mitochondrial protein can be accumulated with no deterioration of bioenergetic integrity. Such accumulation requires the presence of Pi which is accumulated in parallel: the uptake of Ca^{2+} lowers $\Delta\psi$, allowing more protons to be expelled by the respiratory chain (Fig. 8.2). If this were the only process, Ca^{2+} accumulation would soon stop as the major component of Δp was converted from $\Delta\psi$ to ΔpH. In the presence of external Pi, however, the increasing pH causes Pi to enter the matrix via the Pi^-:H^+ symporter (Pi^-/OH^-). This serves two functions: firstly it neutralizes the increase in internal pH, and secondly it allows the accumulated Ca^{2+} to complex with the Pi to form an osmotically inactive calcium phosphate 'gel' (Fig. 8.2).

There is no doubt that mitochondria will respond in the way described above when faced with a cytoplasmic free concentration of Ca^{2+}, $[Ca^{2+}]_c$, in excess of $1 \mu M$, such that the activity of the uniporter exceeds that of the efflux pathway and net accumulation occurs. The activity of the Ca^{2+} uniporter increases as the cube of the free Ca^{2+} in the cytoplasm, and when the $[Ca^{2+}]_c$ is high enough all the respiratory capacity of the mitochondrion can be devoted to the accumulation of the cation. Accumulation will continue until the mitochondrion succeeds in lowering $[Ca^{2+}]_c$ to the 'set-point' at which the rate of uptake and efflux balance. When the mitochondrial matrix contains >10 nmol Ca^{2+} (mg protein)$^{-1}$ in the presence of physiological amounts of Pi, the efflux pathway becomes independent of the matrix Ca^{2+} content since the free Ca^{2+} in the matrix is essentially buffered at a constant value by the formation of the calcium phosphate 'gel'. The 'set-point' is thus independent of matrix Ca^{2+} content above this value and the mitochondria are thus able to act as perfect 'buffers' of $[Ca^{2+}]_c$.

This buffering capacity of the mitochondria has almost certainly evolved in part to serve as a protective mechanism to prevent $[Ca^{2+}]_c$ from rising above a level at which damage to the cell might occur. Cells are very sensitive to $[Ca^{2+}]_c$ levels in excess of $10 \mu M$ and any pathological condition which might generate an abnormally high $[Ca^{2+}]_c$ is thus automatically counteracted. Obviously the mitochondrion has a finite capacity to accumulate any species, and the mitochondrion can only serve as a temporary store of Ca^{2+} until normal conditions resume, when lowering $[Ca^{2+}]_c$ below the mitochondrial set-point, such that the efflux pathway becomes more active than the uniporter, will automatically deplete the matrix of Ca^{2+} which can in turn be pumped out of the cell.

8.3.2 Mitochondrial Ca^{2+} transport and the regulation of mitochondrial metabolism

The previous section has shown that mitochondria can protect the cell under pathological conditions when $[Ca^{2+}]_c$ increases: a remaining question is whether the operation of the transport systems has a physiological role under normal conditions. The activities of three matrix enzymes can be regulated as free Ca^{2+} concentrations in the matrix rise over the range $0.1-1$ μM:

- Pyruvate dehydrogenase phosphatase, which removes Pi from the inactive, phosphorylated form of the pyruvate dehydrogenase complex, thus allowing the v_{max} of the complex to increase.
- NAD^+-linked isocitrate dehydrogenase (ICDH), whose K_m for isocitrate is decreased by Ca^{2+}, allowing a given flux to be achieved at a decreased substrate concentration.
- 2-Oxoglutarate dehydrogenase (OGDH) for which the affinity for substrate is also increased by Ca^{2+} over this concentration range.

The ability of isolated enzymes to be activated by micromolar Ca^{2+} does not in itself show that such a regulatory mechanism exists in the intact cell. However, it is now clear in a number of well-characterized examples, such as the adrenergic stimulation of cardiac muscle contraction, that an increased cytoplasmic $[Ca^{2+}]_c$ is relayed, via the Ca^{2+} transport pathways, into an increase of the mitochondrial matrix $[Ca^{2+}]$ to the range where these key regulatory enzymes would be activated. Enhanced citric acid cycle activity results and hence increased respiration can proceed without any risk of substrate dehydrogenase activity becoming rate limiting or relying on changes in $[ADP]/[ATP]$ or $[NAD]/[NADH]$ ratios to regulate activities. Changes in mitochondrial matrix $[Ca^{2+}]$ can therefore be recognized as providing a mechanism for relaying external signals to the matrix independent of changes in the internal concentrations of metabolites.

It should be emphasized that the two regulatory modes, $[Ca^{2+}]_c$ buffering and matrix enzyme control, are not mutually exclusive: the very subtlety of the mitochondrial Ca^{2+} transport system is that it can perform either function depending on the conditions prevailing in the cell.

8.4 MITOCHONDRIAL METABOLITE (ANION) CARRIERS

The metabolites transported across the inner membrane are predominantly anionic and their transporters are referred to as *anion carriers*. The range of carriers expressed in the inner mitochondrial membrane varies from tissue to tissue. All mitochondria possess the adenine nucleotide and phosphate transporters which are responsible for the uptake of ADP + Pi and the release of ATP to the cytoplasm. Virtually all mitochondria oxidize pyruvate and so possess the pyruvate carrier. However, there is a tissue-specific expression of the other carriers listed in Table 8.1 which correlates with the range of metabolic pathways present in the cell. Thus the liver with its plethora of metabolic pathways has mitochondrial transport pathways for most of the citric acid cycle intermediates, for a number of amino acids and for carnitine and its fatty acyl ester. The more specialized metabolic role played by mitochondria in the heart is reflected in the more restricted variety of carriers, while the mitochondria of a highly specialized tissue such as brown fat can only transport the metabolites acylcarnitine, succinate and pyruvate.

Table 8.1 Metabolite transporters of the inner mitochondrial membrane.

Carrier	Cytoplasm	Function	Matrix	Inhibitors
(a) Adenine nucleotide translocator	ADP^{3-} → ← ATP^{4+}			Atractylate Carboxyatractylate Bongkrekate
(b) Phosphate carrrier	Pi^{-} → ← OH^{-}			Sulphydryl reagents (*N*-ethyl maleimide) (mersalyl)
(c) Dicarboxylate	$malate^{2-}$ → ← Pi^{2-}			Butylmalonate
(d) Tricarboxylate	$citrate^{3-} + H^{+}$ → ← $malate^{2-}$			1,2,3-Benzyl- tricarboxylate
(2) 2-oxoglutarate	2-$oxoglut^{2-}$ → ← $malate^{2-}$			Phenylsuccinate
(f) Glutamate aspartate	$glu^{-} + H^{+}$ → ← asp^{-}			—
(g) Glutamate	glu^{-} → ← OH^{-}			
(h) Pyruvate	pyr^{-} → ← OH^{-}			Cyanohydroxycinnamate
(i) Carnitine	$acylcarnitine^{+}$ → ← $carnitine^{+}$			
(j) Ornithine	$ornithine$ → ← H^{+}			

8.4.1 The ammonium swelling technique for the detection of mitochondrial anion carriers

This technique is of historic interest since it provided the first demonstration of the major metabolite carriers and their functioning as linked systems of antiporters. When non-respiring mitochondria are suspended in an isotonic solution of the ammonium salt of a bilayer-permeant weak acid such as acetate, the mitochondria undergo very rapid osmotic swelling, followed as a decrease in the light scattered as the matrix refractive index decreases to that of the medium (Section 2.5). Swelling can occur because NH_4^{+} crosses as the neutral NH_3 and CH_3COO^{-} crosses as the protonated acetic acid, both passive diffusions, leading to the NH_4^+ and CH_3COO^- concentrations in the matrix becoming equal to those in the suspending medium. There is consequently no charge imbalance or pH imbalance during transport (Fig. 8.3) and massive swelling can occur as the matrix expands to regain osmotic equilibrium with the suspending medium.

In the case of ammonium phosphate, swelling is again observed, indicating that the phosphate anion crosses as the neutral species: which can be visualized as a H^+:$H_2PO_4^-$ symport or the indistinguishable Pi^-/OH^- antiport (Fig. 8.3). When, however, mitochondria are suspended in isotonic ammonium malate no swelling occurs until a low concentration of Pi is added. This is required since malate permeates the inner membrane on a carrier protein in exchange for Pi. Phosphate can thus cycle across the membrane between the malate and phosphate carriers. The situation with a tricarboxylic acid such as citrate is still more complex: low concentrations of both phosphate and malate are required, since the citrate/isocitrate carrier exchanges with malate which in turn exchanges with Pi (Fig. 8.3).

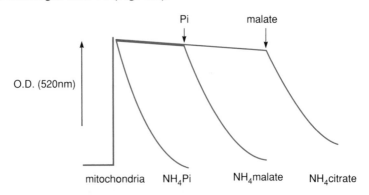

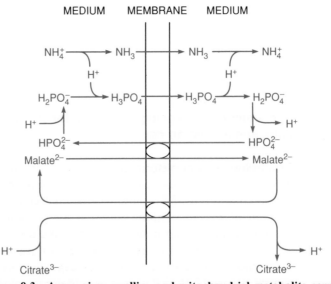

Figure 8.3 Ammonium swelling and mitochondrial metabolite carriers.

Careful examination of the stoichiometries in Fig. 8.3 reveals that the net accumulation of $H_2PO_4^-$ is accompanied by the entry of one proton; that malate uptake is accompanied by two protons; and that three protons enter with each citrate. In thermodynamic terms this means that the phosphate gradient in/out can in theory equal the proton concentration gradient out/in (membrane potential has no effect), that the malate gradient can equal the square of the proton concentration gradient and that citrate or isocitrate can be accumulated up to the cube of the proton concentration gradient.

8.4.2 The structure and function of mitochondrial anion carriers

Table 8.1 lists the major anion carriers which have now been identified in the inner membrane. Since the neutral species of a weak acid such as acetate can permeate through lipid bilayer regions of the membrane without the need for a carrier, how does one establish that a specific carrier protein is required for transport? In the case of the polycarboxylic acids such as malate and citrate the issue is clear-cut since the fully protonated form is present in only a tiny proportion and would in all likelihood still be too hydrophilic to permeate without a carrier. However, the firm identification of a carrier for pyruvate is less intuitive and is based on the existence of a specific inhibitor (see Section 8.5.7).

The four carriers whose primary sequences have been determined at the time of writing are the phosphate carrier, the adenine nucleotide carrier, the oxoglutarate/malate carrier and the brown fat uncoupling protein (which can be considered as an anion carrier since it can transport OH^- or Cl^- by uniport and it shows close homology to the adenine nucleotide carrier, Klingenberg, 1990). Each protein is about 30 kDa, formed from three closely related 100 amino acid repeats, each of which may fold into two transmembrane α-helices, and appears to have originated from a common ancestor gene (Runswick *et al.*, 1990).

In addition to these four carriers, the dicarboxylate carrier, aspartate/glutamate carrier, pyruvate carrier, tricarboxylate carrier and carnitine carrier have all been purified and functionally reconstituted into liposomes. Although the primary sequence of these carriers has yet to be determined, most share a M_r of 30 000 kDa, suggesting that they might also belong to this carrier 'superfamily'.

8.4.3 The adenine nucleotide carrier

Review Klingenberg 1989

In bacteria and chloroplasts the ATP synthase produces the ATP in the same compartment in which it is utilized, but mitochondria synthesize ATP in the matrix and then export the nucleotide to the cytosol. Two carriers are involved: the phosphate carrier for the uptake of Pi, and the adenine nucleotide carrier for the uptake of ADP and export of ATP (Fig. 8.4).

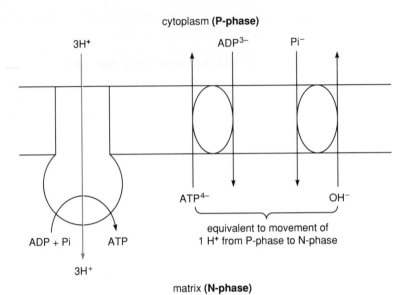

Figure 8.4   The uptake of ADP and Pi and the export of ATP from mitochondria involves an additional translocated proton.

The adenine nucleotide carrier catalyses the 1:1 exchange of ADP^{3-} for ATP^{4-} across the inner membrane (Fig. 8.4); it does not transport AMP. The total pool size of adenine nucleotides in the matrix (ATP + ADP + AMP) does not change, as the uptake of a cytosolic nucleotide is automatically compensated by the efflux of a nucleotide. Even if mitochondria are suspended in a nucleotide-free medium (as they are during preparation), no loss of nucleotide normally occurs.

A number of inhibitors are specific for the translocator. Atractyloside, a glucoside isolated from the Mediterranean thistle *Atractylis gummifera*, is a competitive inhibitor of adenine nucleotide binding and transport. The closely related carboxyatractylate binds more firmly (K_d 10^{-8} M) and cannot be displaced by adenine nucleotides. Bongkrekic acid is produced by *Pseudomonas cocvenenans* and derives its name from its discovery as a toxin in contaminated samples of the coconut food product bongkrek. It is an uncompetitive inhibitor of the translocator.

The polypeptide of the adenine nucleotide carrier was initially identified by its ability to bind [^{35}S]carboxyatractylate, binding which is sufficiently stable to survive solubilization of the membrane with the non-ionic detergent Triton X100 and subsequent purification. A second approach involved the use of photoaffinity analogues of adenine nucleotides, which are competitive inhibitors of transport and bind to the catalytic site of the carrier in the dark. In UV light they lose N_2 to form highly reactive nitrene-free radicals which attach covalently to the nearest amino acids of the polypeptide to which they are bound, i.e. the translocator. The primary sequence was initially determined by peptide sequencing, and hydrophobicity plots suggest that the 30-kDa peptide might have six transmembrane helices.

In the model of the translocator proposed by Klingenberg, the carrier would function as a 60-kDa dimer with just one nucleotide binding site. The transport path would follow the central two-fold axis. The dimer would exist in two states differing in the orientation of this single nucleotide binding site (Fig. 8.5) which can either be accessible from the cytosol (C-state), and bind carboxyatractylate (one molecule per dimer), or from the matrix (M-state), and bind bongkrekic acid. As these two inhibitors have completely different structures it follows that the properties of the nucleotide binding site must differ according to the orientation of the site. The two inhibitors bind very tightly so that the carrier can be fixed in a C or an M state. Binding of adenine nucleotide is less tight because, it is argued, intrinsic binding energy is realized to drive conformational changes in the relatively fluid dimeric protein in order to facilitate transport. These changes do not occur following binding of the inhibitors; these molecules are not transported.

Note that the model (Fig. 8.5) is a 'rocking bananas' model where slight conformational changes make the substrate binding site alternately accessible from either side of the membrane. Direct experimental evidence for such a conformational change came from monitoring the accessibility of lysine residues to the hydrophilic reagent pyridoxyl-5-phosphate when added to either the cytoplasmic or matrix face of the membrane: distinct sets of residues were exposed in the M- and C-conformations.

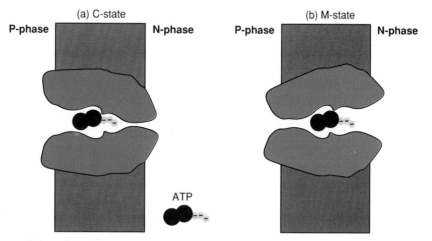

Figure 8.5   The adenine nucleotide translocator.
The translocator undergoes a conformational change between the C-state, in which the nucleotide binding site faces the cytoplasm (P-phase) and the translocator binds the inhibitor carboxyatractylate, and the M-state, where the binding site is exposed to the matrix (N-phase) and the translocator binds bongkrekic acid. This type of conformational change is known as a 'rocking bananas' mechanism.

8.4.4 **The phosphate carrier**

Selected reference Ferreira *et al.* 1989

The phosphate carrier catalyses the electroneutral transport of $H_2PO_4^-$, either in exchange for OH^- or by symport with a proton, the two being indistinguishable. The carrier is inhibited by mercurial reagents such as *p*-mercuribenzoate and mersalyl, and also by *N*-ethylmaleimide, although none of the inhibitors is completely specific. The phosphate carrier is extremely active. Because of the proton symport, the distribution of Pi across the membrane is influenced by ΔpH, and a factor complicating the measurement of H^+/O ratios in mitochondria by oxygen pulse methods is the facility with which Pi can redistribute across the membrane and partially neutralize any ΔpH generated by respiration (Section 4.4).

8.4.5 **Interrelations between the ATP synthase, phosphate carrier and adenine nucleotide carrier**

The complete system for mitochondrial ATP synthesis and export requires the ATP synthase, adenine nucleotide carrier and phosphate carrier. While the translocator transports ADP and ATP symmetrically when there is no membrane potential, under normal respiring conditions uptake of ADP and efflux of ATP are preferred, corresponding to the physiological direction of the exchange. The reason for this asymmetry lies with the relative charges on the two nucleotides. ATP is transported as ATP^{4-}; ADP is transported as ADP^{3-}. The resulting charge imbalance means that the equilibrium of the exchange is displaced ten-fold for each 60 mV of membrane potential. The combined effect of the phosphate carrier and adenine nucleotide carrier is to cause the influx of one additional proton per ATP synthesized (Fig. 8.4). Note that although the additional proton apparently enters with Pi this is an electroneutral process and the charge of the additional proton is used to drive the exchange of ADP^{3-} for ATP^{4-}.

The thermodynamic consequences of this are considerable. First, up to one-third of the Gibbs energy of the cytosolic ATP/ADP + Pi pool comes, not from the ATP synthase itself, but from the subsequent transport. Secondly, as four protons appear to be used to synthesize a cytosolic ATP but only three for a matrix ATP, it follows that in state 4 the cytosolic ΔG_p (Section 3.2) can be up to 33% higher than in the matrix, or that produced by inverted SMPs. This can be observed: isolated mitochondria can maintain a ΔG_p of up to 66 kJ mol^{-1} in contrast to a value of less than 50 kJ mol^{-1} for SMPs (see Table 4.1).

8.4.6 **The carnitine carrier**

Fatty acids are activated to acyl CoA on the outer mitochondrial membrane and converted to acylcarnitines via carnitine *N*-acyltransferase I. The acylcarnitines are hydrophilic and cross into the mitochondrial matrix on the carnitine carrier in a 1:1 exchange for carnitine. In the matrix, carnitine N-acyltransferase II regenerates acyl CoA for β-oxidation and liberates the carnitine which is required for the exchanger. The carrier is not specific for the acyl chain length and will transport acetylcarnitine and short- and long-chain acylcarnitines.

8.4.7 The dicarboxylate and tricarboxylate carriers

The dicarboxylate and tricarboxylate carriers are highly active in liver mitochondria but almost absent from heart. Both allow the net export of citric acid cycle intermediates from the matrix to be used for gluconeogenesis and fatty acid synthesis respectively. The dicarboxylate carrier catalyses the electroneutral exchange of malate^{2-} or succinate^{2-} for HPO_4^{2-} and is inhibited by butylmalonate. The tricarboxylate carrier exchanges citrate^{3-} or isocitrate^{3-} for malate^{2-} but is also electroneutral since it cotransports a proton together with citrate. The carrier can be inhibited by 1,2,3-benzyltricarboxylate.

8.4.8 The transfer of electrons from cytoplasmic NADH to the respiratory chain

The NADH which is produced in the cytoplasm of mammalian cells, for example by glycolysis, does not have direct access to complex I, whose NADH binding site is located on the inner face of the inner membrane. The inner membrane is impermeable to NADH and, therefore, two strategies for oxidation of cytosolic NADH are employed (Fig. 8.6):

(a) The malate–aspartate shuttle. The 2-oxoglutarate and glutamate–aspartate carriers (Table 8.1) occur in many mitochondria. One function of these carriers is as components of the malate–aspartate shuttle (Fig. 8.6), a device allowing the oxidation of cytosolic NADH by the respiratory chain. A thermodynamic problem posed by this process is that the E_h of the cytosolic $NAD^+/NADH$ couple is considerably higher (i.e. less reducing) than the equivalent matrix couple. This thermodynamic impasse is overcome by the electrical inbalance of the glutamate–asparate carrier, which exchanges glutamate$^-$ plus a proton for aspatate$^-$ and is therefore driven in the direction of glutamate uptake and aspartate expulsion when a $\Delta\psi$ exists across the membrane.

(b) The s,n-glycerophosphate shuttle provides a second means for the oxidation of cytosolic NADH (Fig. 8.6). This makes use of the two s,n-glycerophosphate dehydrogenases present in most cells, a cytosolic enzyme coupled to NAD^+ and an enzyme on the outer face of the inner mitochondrial membrane feeding electrons directly to UQ. In this case the

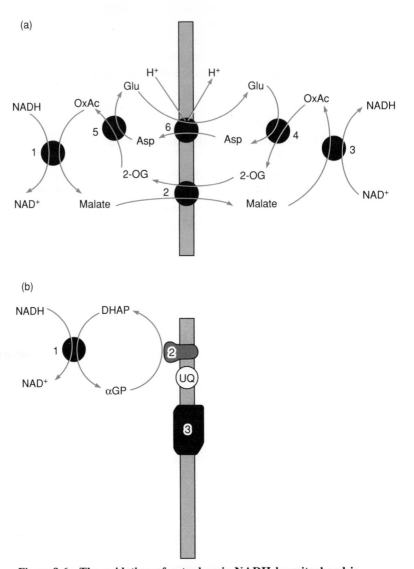

Figure 8.6 The oxidation of cytoplasmic NADH by mitochondria.
(a) The glutamate–aspartate shuttle: (1) cytoplasmic NADH is oxidized by cytoplasmic malate dehydrogenase; (2) malate enters matrix in exchange for 2-oxoglutarate; (3) malate is reoxidized in the matrix by malate dehydrogenase, generating matrix NADH; (4) matrix oxaloacetate trans-aminates with glutamate to form aspartate and 2-oxoglutarate (which exchanges out of the matrix); (5) 2-oxoglutarate transaminates in the cytoplasm with transported aspartate to regenerate cytoplasmic oxaloace-tate and to give cytoplasmic glutamate which (6) re-enters the matrix by proton symport in exchange with aspartate. (b) The *sn*-glycerophosphate shuttle for the oxidation of cytoplasmic NADH: (1) cytoplasmic *sn*-glycerophosphate dehydrogenase (NAD^+-linked), (2) inner membrane *sn*-glycerophosphate dehydrogenase (flavoprotein-linked), (3) complex III.

directionality is induced by feeding electrons to the quinone pool at a potential close to 0 mV.

8.4.9 The glutamate, pyruvate and ornithine carriers

The glutamate$^-$/OH$^-$ exchanger is present in liver and kidney mitochondria and provides an alternative to the glutamate–aspartate carrier for the transport of glutamate. Because pyruvate is a monocarboxylic acid, it could be argued that it would cross bilayer regions without the need for a carrier, following the precedent of acetate. However, cyanohydroxycinnamate inhibits pyruvate transport specifically, indicating that a carrier protein is involved, and this has been confirmed by the subsequent purification and functional reconstitution of the carrier. Kidney mitochondria transport ornithine in a 1:1 electrical exchange of neutral ornithine with H$_2$PO$_4^-$.

8.5 BACTERIAL TRANSPORT

Reviews   Booth 1988, Kornberg and Henderson 1990

Bacteria survive in environments which are far more variable, and usually more hostile, than anything experienced by a mitochondrion or chloroplast. As a consequence they have developed a variety of mechanisms for the transport of metabolites, such as amino acids and sugars, from the external medium where such molecules occur at very low concentrations to the cytoplasm where the concentrations must be considerably higher to sustain metabolism. The origins of the chemiosmotic theory lay in Mitchell's desire to explain such 'active transport' in bacteria, and there is now extensive evidence that many such transport processes are directly driven by Δp. However, there are also active transport processes that are powered by a Na$^+$ electrochemical gradient or by the direct hydrolysis of ATP, while a fourth class relies upon phosphoenolpyruvate as immediate energy source. Finally there are examples of anion exchange reactions, e.g. hexosephosphate/phosphate antiports. We shall consider each of these in turn.

8.5.1 Proton symport systems

Reviews   Henderson 1990, 1991, Kaback, 1990, Kaback et al. 1990, Wright et al. 1986

The first evidence that active transport across a bacterial cytoplasmic membrane could be linked to Δp was obtained for lactose uptake by E. coli. This was known to involve the product of the lac Y gene which is traditionally known as the lactose (lac) permease, but 'transporter' is to be preferred to 'permease', and whose expression is inducible. The key experimental observation was that addition of lactose to an anaerobic and lightly buffered suspension of cells in which *lac Y* expression was induced resulted in the pH of the external medium showing a small alkaline shift. The pH change was not seen if *lac Y* expression was not induced. These experiments, which were analogous to the oxygen pulse experiment for detecting respiration-driven proton translocation (Sections 4.3 and 4.4), thus indicated that lactose entered the cells in symport with one or more protons. Quantitation is difficult but all available evidence is that the stoichiometry is $1H^+$/lactose. As lactose is un-

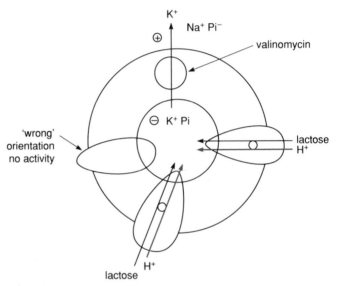

Figure 8.7 Reconstitution of the lactose: proton symporter from *E. coli*. The *lac* transporter (permease) is purified from cells in which it is overexpressed by selective solubilization of cytoplasmic membranes with octylglucoside followed by ion-exchange chromatography. Identification of the transporter throughout the original purification was aided by covalent attachment of a radiolabelled and highly specific photoaffinity label. Unilamellar vesicles (sometimes called proteoliposomes) containing transporter polypeptide are prepared by dilution of octylglucoside in the presence of phospholipids, freezing and subsequent sonication. Uptake of lactose can be studied by incubation of the vesicles in 50 mM potassium phosphate (pH 7.5), addition of valinomycin and then dilution into sodium phosphate, pH 7.5, containing [^{14}C] lactose. A $\Delta\psi$, negative inside is thus generated and the resulting uptake of lactose can be determined at various times by rapid dilution of the vesicles followed by immediate filtration. The radioactivity retained within the vesicles trapped on the filter can then be determined. Such experiments showed that the turnover number and K_m for lactose were similar to the values observed with cytoplasmic membranes, establishing that the purified protein was in an unperturbed state. In a variation of the experiment $^{86}Rb^+$ can be included in the lumen of the vesicles and its efflux, which is stimulated by the presence of lactose, monitored in parallel with the uptake of [^{14}C] lactose. Such studies indicate the movement of one Rb^+ per lactose, consistent, for reasons of charge balance, with a stoichiometry of one proton moving per lactose.

charged this means that the lactose$_{in}$/lactose$_{out}$ gradient at equilibrium is maintained by Δp and can attain a value of about 3000 at the usual values of Δp.

The purified *lac* transporter protein has been inserted in phospholipid bilayers and shown to catalyse the expected symport. Such experiments demonstrated that the single polypeptide chain was sufficient for the movement of both the H$^+$ and the lactose. To drive accumulation an artificial Δp (interior alkaline and negative) is generated across the vesicle membrane by transferring a suspension of vesicles held in the presence of valinomycin and high K$^+$ into a medium at slightly more acidic pH containing radiolabelled lactose and low K$^+$ (Fig. 8.7). A Δp of the correct polarity will be instantaneously established (K$^+$ efflux generates the $\Delta\psi$) and the uptake of lactose can be assessed by rapid filtration of the vesicles. In principle this experiment would work best if all the lactose:proton symporter molecules were to insert into the vesicles with the same orientation as they have in intact cells. In practice uniform orientation can be very difficult to achieve, but in this case carriers with the opposite orientation will be inert because they are not exposed to the appropriate Δp. Lactose:proton symporter activity has also been reconstituted in vesicles also containing cytochrome *o* from *E. coli* (Section 5.13.2) which generates Δp by oxidizing ubiquinol.

The DNA-derived amino acid sequence of the lactose:proton symporter is consistent with 12 transmembrane α-helices (Fig. 8.8). This predicted structure indicates the presence of some hydrophilic side chains within these helices, and it is tempting to propose that such residues may contribute to the passage of the proton through the membrane (Fig. 8.8). Work with bacteriorhodopsin (Section 6.5) has shown that site-directed mutagenesis of such candidate residues can give clues about mechanism but that the three-dimensional structure is needed to substantiate mechanistic proposals. The lactose symporter will require a structure, together with mutagenesis studies and kinetic measurements, in order to describe the pathways for the proton and the sugar through the protein.

In *E. coli* a range of other sugars, including arabinose, xylose and galactose, are transported in symport with protons, while other bacterial species will have comparable systems. Surprisingly, although the proteins from these three transport systems of *E. coli* resemble each other in sequence, they differ from the lactose symporter very significantly, with the exception that they too are proposed to form 12 transmembrane helices. The sequences of the arabinose, xylose and galactose symporters have a remarkable similarity to several sugar transport proteins from eukaryotic cells, for example the glucose transporter from the cytoplasmic membranes of liver cells. The latter is a passive transporter and thus an intriguing challenge now is to identify the features of the bacterial proteins that confer on them the capacity to transport the sugars in symport with a proton. Attempts to convert by site-directed mutagenesis one of the passive transporter proteins into the active type have so far been unsuccessful, but alteration of Glu-325 to Ala in the *lac* transporter does give a protein that catalyses downhill lactose movement without proton translocation.

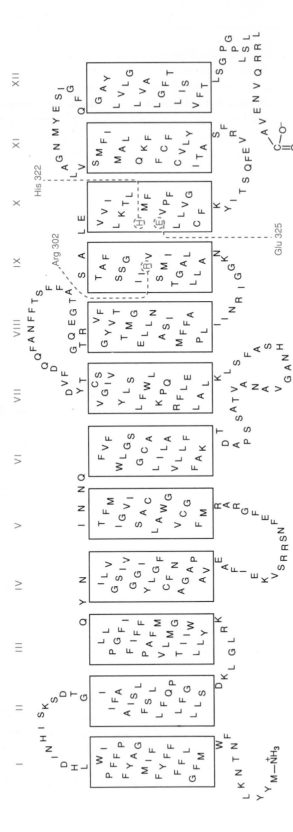

Figure 8.8 The transmembrane structure of the lactose: proton symporter.

A 12 α-helix model for the secondary structure of the *lac* transporter based on the hydropathy profile of the amino acid sequence (using single letter amino acid code) and supported by circular dichroism studies (indicating high α-helical content), antibody binding, fusions of the *lac Y* gene with the gene (*phoA*) for a periplasmic alkaline phosphatase and limited proteolysis studies (all indications of the locations of non-helical stretches of polypeptide). Alteration of His-322, Glu-325 and Arg-302 to certain other residues via site-directed mutagenesis yielded transporter defective in transloca- tion reactions (i.e. accumulation of lactose against its concentration gradient) involving protonation or deprotonation events. Model building sug- gested that the interplay of these residues on helices IX (302) and X (322 and 325) could provide a proton relay system (Kaback *et al.* 1990a) analogous to the charge relay system in chymotrypsin. However, the general difficulty of interpretation of site-directed mutagenesis studies is illustrated by the finding that other mutations at His-322 (e.g. to Phe or Asn) do not abolish co-transport of lactose and protons (Franco and Brooker, 1991). It is nevertheless difficult to exclude the possibility that His-322 is an integral part of a H^+-translocation pathway but that it can be bypassed in certain mutants. The properties of several mutants with certain amino acid substitutions indicate that the H^+ and lactose are transported along the same pathway (e.g. a single-channel-like structure) within the protein. The residues identified by mutation as functionally significant lie on the six helices V–X which may form a channel with the side chains of these residues facing inward (Brooker, 1990).

The components of Δp can in principle drive the uptake of any type of nutrient molecule (Table 8.1). In the case of a positively charged molecule, for example lysine at neutral pH, a possible mechanism would be a uniport driven by $\Delta\psi$ and the logarithm of the equilibrium accumulation ratio would be a function of $\Delta\psi$. If a proton were to be co-transported with lysine then both ΔpH and $\Delta\psi$ would contribute to the driving force and the equilibrium distribution would be a function of (ΔpH + 2 $\Delta\psi$). A monoanionic species in symport with one proton would be driven only by ΔpH. However, as Δp in most species of bacteria is usually predominantly present as $\Delta\psi$ such a symport is improbable. On the other hand, symport of an anion with two protons would allow an equilibrium to be attained which would be proportional to (2 ΔpH + $\Delta\psi$). These should be seen as illustrations because there is no *a priori* reason why other proton stoichiometries should be excluded.

8.5.2 Sodium symport systems

Reviews Dibrov 1991, Maloy, 1990

Protons were once thought to be the only cations to move in symport with sugars and amino acids into *E. coli*. It is now known that melibiose and proline transport both occur in symport with Na^+, driven by the Na^+ electrochemical gradient $\Delta\tilde{\mu}_{Na^+}$ (note that the term 'sodium motive force' is not in general use). Unlike with *P. modestum* (section 5.13.5) there is no primary Na^+ pump. Instead $\Delta\tilde{\mu}_{Na^+}$ is maintained by an electroneutral Na^+/H^+ exchanger in the cytoplasmic membrane which equilibrates the Na^+ and H^+ concentration gradients. $\Delta\psi$ is, of course, a delocalized parameter and so the $\Delta\psi$ generated by proton pumping will be a component of $\Delta\tilde{\mu}_{Na^+}$.

There are now many instances where the role of a Na^+ circuit has been established for various species of bacteria. In some cases, e.g. active transport into alkaliphilic bacteria (Section 4.9.2d), this can be rationalized on the basis that Δp is too small to drive active transport. It is not clear why *E. coli* should use both H^+ and Na^+ symports, but nor is it appreciated why methane synthesis (Section 5.13.4) seemingly involves both Na^+ and H^+ translocation. The interplay of Na^+ and H^+ movements across the bacterial cytoplasmic membrane requires further investigation.

8.5.3 Transport driven directly by ATP hydrolysis

Reviews Ames 1990, Epstein 1990, Higgins *et al.* 1990

This mode of transport will be illustrated by reference to two types of system.

The concentration of K^+ in the bacterial cytoplasm is generally much higher than in the surrounding medium. The gradient of K^+ has to be actively maintained, and its mechanism has been studied in depth for *E. coli*. Doubtless related systems operate in other genera.

One of the K^+ uptake systems in *E. coli* is known as Kdp and is induced when the external K^+ is very low. Three gene products, KdpA, KdpB and

KdpC, have been identified as having characteristics of integral membrane proteins and together they constitute a K^+-dependent ATPase. The KdpA protein is thought to be responsible for the initial binding of K^+ at the periplasmic surface. The site of ATP hydrolysis resides on the KdpB protein; an acyl-phosphate intermediate forms, in contrast to the $F_1.F_o$ ATP synthase, but similar to mammalian $E_1.E_2$ ion pumps such as the (Na^+/K^+)-ATPase. Indeed, there are significant regions of sequence similarity betwen the latter and the KdpB protein. It is not known whether the Kdp system catalyses electrogenic import of K^+ or whether K^+ entry might be in exchange for another ion.

A second major K^+ transport system in *E. coli* is the constitutive TrkA system. This has a lower affinity than Kdp and its energetics have not been fully elucidated. It appears to require both ATP hydrolysis and the presence of a protonmotive force for activity but as the latter drives the synthesis of ATP it is conceivable that Δp is not directly involved. In principle K^+ uptake could be via a uniport in response to $\Delta\psi$ but this seems not to be a mechanism that has been adopted.

A major group of transport systems in Gram-negative bacteria involve water-soluble proteins in the periplasm that first bind the transport substrate after it enters from the external medium through the porin of the outer membrane. The substrates handled in this way, at least for enteric bacteria such as *E. coli* and *Salmonella typhimurium*, include Pi, SO_4^{2-}, ribose, maltose and histidine. The periplasmic binding proteins have a high affinity for their substrates; their specificity is known from a series of high-resolution X-ray diffraction structures to be conferred by a set of hydrogen bonds which interact with the substrate. If the contents of the periplasm are released, for example by exposing the cells to an osmotic shock, transport is greatly inhibited.

Genetic analysis of many of these transport systems has shown that three or four other polypeptides are involved in addition to the periplasmic binding protein. The histidine and maltose uptake systems serve as examples (Fig. 8.9). Two of these additional polypeptides, known as M and Q for histidine and F and G for maltose, are membrane spanning on the basis of their high content of hydrophobic amino acids and appear to recognize a substrate even in the absence of the periplasmic binding protein. Thus a mutant lacking the periplasmic binding protein, but carrying a secondary mutation, still transports maltose, albeit with very low affinity.

A third protein P (histidine) or K (maltose) has a sequence consistent with its attachment via Q (F) and/or P (G) to the inner surface of the cytoplasmic membrane, an organization supported by combined genetical and biochemical experiments. P and related polypeptides of other systems have regions of sequence, including the Gly-X-X-X-X-Gly-Lys- motif seen also in F_1. F_o ATP synthases (Section 7.6.2b), that indicate the presence of an ATP-binding motif or cassette (hence the proteins are considered members of the ABC family of transporters but F_1. F_o is otherwise distinct and does not belong to the family). P-type subunits can be labelled by ATP analogues, further suggesting that transport might be driven by ATP hydrolysis. This has been confirmed with

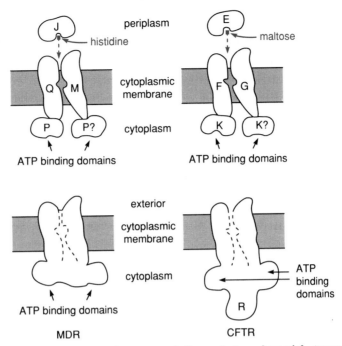

Figure 8.9   Diagrammatic representation of two bacterial transport systems, for histidine in *S. typhimurium* and maltose in *E. coli*, that involve periplasmic binding proteins and comparison with two related proteins from mammalian systems.

Both the histidine and maltose systems involve four polypeptides and it is speculated that the polypeptide responsible for hydrolysis of ATP (P in the histidine system and K in the maltose transporter) is probably present in two copies for each copy of the other subunits. Analysis of other systems e.g. that for ribose in *E. coli*, shows that the subunit equivalent to P or K is larger and contains two ATP binding domains. The multidrug resistance (MDR) protein and the product of the gene in which a defect gives rise to cystic fibrosis, known as the CFTR protein, have comparable ATP binding domains to the bacterial transport systems but are made up of a single polypeptide, approximately equivalent to fusing two copies of P with one each of Q and M of the histidine system, and lack an equivalent of the periplasmic binding proteins. CFTR also has a fifth domain known as R which may have a regulatory role.

a reconstituted histidine transport system and also with a strain of *E. coli* in which ATP hydrolysis with a stoichiometry of about 2ATP/molecule transported could be demonstrated to be associated with transport of maltose. This high stoichiometry (remember that two ATP are equivalent to about six protons translocated) could account for the very high accumulation ratios, up to 10^5, that are achieved by these transport systems. Indeed, if a dianion, e.g. sulphate or phosphate at pH values above 7, were to be taken up by a chemiosmotic mechanism then up to four protons would need to move in symport at a cytoplasmic membrane potential of 180 mV to achieve accumulation ratios of this order.

Thermodynamic considerations do not account for the occurrence of periplasmic binding proteins in the ATP-dependent transport systems and their absence from proton symport systems. Any advantage to be gained from the avid binding of the substrate would be balanced by the need of the membrane-bound components to abstract the substrate from its binding site. This is an unresolved issue, but a factor to be taken into consideration is that analogous ATP-dependent transport systems have recently been identified in Gram-positive organisms that do not have a periplasm. In these instances the equivalent to the periplasmic binding protein appears to be anchored to the cytoplasmic membrane. By analogy it seems likely that in Gram negative organisms the binding protein functions to deliver the bound substrate to the membrane-spanning components of the system.

Interest in the bacterial ATP binding and periplasmic binding protein-dependent transport systems has been significantly enhanced by the finding that there are considerable similarities with two mammalian proteins (Higgins, 1992). Prominent amongst these is the product of the gene in which mutation leads to cystic fibrosis. The gene product is called the CFTR (cystic fibrosis transmembrane regulator) protein but at the time of writing its physiological function has not been elucidated. The comparison with the bacterial proteins would suggest that it is part of an ATP-dependent transport system, but much evidence suggests that it is a chloride channel. Conceivably the ATP binding site could be present to permit a regulatory role, perhaps analogous to that exerted by GTP and GDP binding to G-proteins. The second mammalian system is the multidrug resistance (MDR) protein which is an ATP-dependent system of low specificity that exports drugs from cells and thus causes problems in certain drug treatments e.g. for cancer. The connection between this protein and the CFTR protein has recently been strengthened by the finding that the MDR protein can also act as a chloride channel that is dependent on the binding but not on the hydrolysis of ATP (Valverde *et al.*, 1992). To complete the connection between the bacterial and mammalian systems, there is recent evidence that there is an analogue of mdr in a *Streptomyces* species that produces antibiotics to which it is resistant (Guilfoile and Hutchinson, 1991).

8.5.4 Transport driven by phosphoryl transfer from phosphoenolpyruvate

Reviews   Meadow *et al.* 1991, Lengeler 1990

The fourth general class of transport mechanism is the phosphotransferase system (often abbreviated PTS) which catalyses the transport of several hexose and hexitol sugars, e.g. glucose and mannitol, in many different bacterial genera including *E. coli* and *Staphylococcus aureus*. A distinctive feature of this system (Fig. 8.10) is that an integral membrane polypeptide, specific for a particular sugar, and generally called enzyme II, binds the sugar from the external surface and phosphorylates it to give, for example, glucose 6-phosphate, which is released into the cytoplasm. This has often been called a

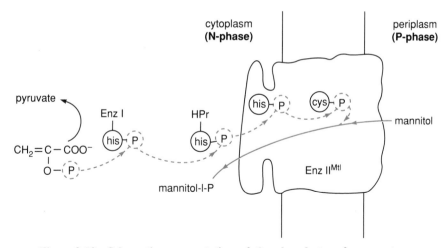

Figure 8.10 **Schematic representation of the phosphotransferase system (PTS) for mannitol in *E. coli*.**
In this case the mannitol-specific enzyme II ^{mtl} accepts phosphate onto a histidine residue from which it moves to a cysteine residue (a thus far unique example of a phosphocysteine catalytic intermediate in biology) before transfer to the incoming mannitol. As mentioned in the text, in other PTS transport systems, e.g. for glucose in *E. coli*, separate enzyme III catalyses transfer of phosphate from HPr to the enzyme II ^{glc}. Sequence analysis of the enzyme II ^{mtl} shows that a cytoplasmic domain is in effect a fused type III enzyme. The N-terminal region contains the mannitol-binding domain. The enzyme II molecules (excluding where appropriate fused enzyme III domains) all have of the order of 650 amino acids, between 350 and 380 of which could fold to form transmembrane α-helices. There may be extensive regions of relatively hydrophilic polypeptide that extend from the bilayer. Such proteins must presumably fold to give a hydrophilic central channel in which phosphorylation occurs (Lengeler, 1990).

group translocation mechanism because substrates and products have to approach and leave the catalytic site of enzyme II along defined pathways (i.e. glucose should not have access to this site from the cytoplasm).

The ultimate source of the phosphate group for the phosphorylation is cytoplasmic phosphoenolpyruvate (PEP), which first transfers its phosphate to the N-1 nitrogen of a histidine residue in a soluble protein known as H-Pr. This transfer is catalysed by enzyme I, a water-soluble cytoplasmic protein that has no sugar specificity. Subsequently the phosphate group is transferred from H-Pr to a histidine residue on enzyme II, either by the catalytic activity of enzyme II itself or as a result of the activity of yet another enzyme known as enzyme III. It is known from crystallographic studies on soluble proteins that the incorporation of a phosphate group can cause conformational changes. Thus it is conceivable that phosphorylation of enzyme II could lead to a key conformational change that would prevent a sugar bound from the external media from returning there.

Under physiological conditions the ΔG for PEP hydrolysis is more negative than for ATP hydrolysis and is sufficient for phospho transfer to histidine. The driving force for transport is considerable; it has been calculated that at equal concentrations of pyruvate and PEP the equilibrium intracellular concentration of a phosphorylated sugar would reach approximately 100 M at 10^{-6} M external sugar. Such equilibration is apparently not achieved, indicating the requirement for tight control of these transport systems. Another energetic facet is that the bacterial cell expends one PEP molecule for the acquisition of an intracellular phosphorylated sugar. Other active transport processes (e.g. energy-consuming proton symports) are followed by intracellular phosphorylation at the expense of ATP. The conversion of PEP to pyruvate in the pyruvate kinase reaction yields only one molecule of ATP, in a reaction with a large negative Gibbs energy change under cellular conditions, and thus it is energetically advantageous to harness more fully the energy associated with PEP hydrolysis to drive both the transport and phosphorylation events.

8.5.5 Transport driven by anion exchange

Reviews   Ambudkar and Rosen 1990, Maloney 1990

There is evidence, especially for the Gram-positive organism *Streptococcus lactis*, that transport can occur by anion-exchange systems analogous, at least in principle, to those found in the inner mitochondrial membrane (Section 8.4). The system identified is concerned with the uptake of glucose 6-phosphate (G6P) into the cell. In one mode two $G6P^-$ anions move into the cell in exchange for the export of one $G6P^{2-}$. This apparently curious exchange is therefore electroneutral and is equivalent to the net entry of one $G6P^{2-}$ and two H^+ and so is thermodynamically equivalent to a $2H^+$:$G6P^{2-}$ symport with ΔpH acting as the sole driving force. The exchange system can also operate by moving two $H_2PO_4^-$ anions out and one $G6P^{2-}$ in. The exchange is electroneutral and the advantage to the cell is that growth on G6P provides too little carbon and too much phosphorus and therefore extrusion of excess phosphate in exchange for G6P is favourable. As presently understood these exchange systems do not involve electrogenic movement of molecules across membranes.

A different type of exchange system that generates Δp has recently been described. In the anaerobe *Oxalobacter formigenes* oxalate is taken up as a dianion. Once in the cell, it is decarboxylated:

$$(COO^-)_2 + H^+ \rightleftharpoons HCOO^- + CO_2$$

Formate as a monoanion exits from the cell in exchange for the oxalate, with the overall effect that the exchange is responsible for generating a membrane potential, positive outside (Fig. 8.11). There is also a tendency for a ΔpH to develop, alkaline inside, owing to the consumption of a proton during the decarboxylation reaction. This mechanism requires that the CO_2 leaves the cell

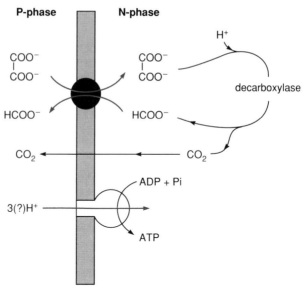

P-phase   **N-phase**

Figure 8.11   Generation of Δp by anion exchange in Oxalobacter formigenes.
The cycle of influx, decarboxylation and efflux effectively constitutes a proton pump with a stoichiometry of one proton per turnover. If the H^+/ATP ratio for the ATP synthase is 3 then the maximum stoichiometry of ATP synthesis would be 1 for each 3 oxalate molecules decarboxylated.

in the gaseous form. If it were to be hydrated to HCO_3^- and exported as such from the cell no $\Delta\psi$ would be generated. In line with this requirement it would be expected that carbonic anhydrase activity is very low or absent in these cells.

Descriptions of other kinds of ion-exchange reactions across the bacterial cytoplasmic membrane are given in Ambudkar and Rosen, (1990).

8.6 TRANSPORT (MOVEMENT) OF BACTERIAL CELLS

Review   Armitage 1992

Many, but by no means all, species of bacteria are motile, usually as a consequence of the rotation of one or more helical flagella that extend from the surface of the cell (Fig. 8.12). The basal body of a flagellum is embedded in the cytoplasmic membrane, traverses the periplasm, the peptidoglycan and outer membrane and connects to a filament that extends into the external phase. Movement of the filament is driven by a 'motor' embedded in the cytoplasmic membrane. The motor rotates at up to 3000 rev/min about an axis perpendicular to the plane of the membrane and is usually driven by the protonmotive force, although a Na^+ electrochemical gradient is used in some organisms where Na^+ circuits are generally important, e.g. *Vibrio* species. The molecular mechanism of torque generation is unknown, although in its simplest terms protonation of (or binding of Na^+ ions to) a rotor component could cause

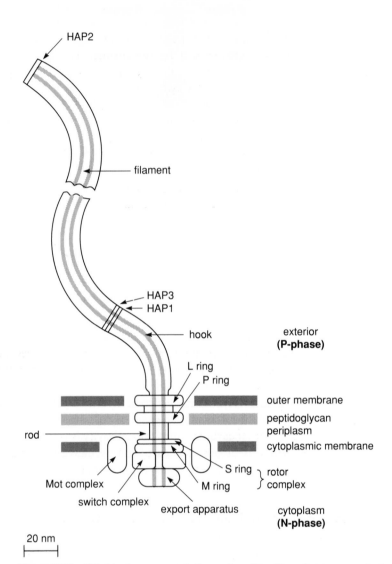

20 nm

Figure 8.12 Schematic representation of a flagellum from an enteric bacterium.
The Mot complex, probably forming an 8–12-subunit circlet around the motor, is involved in transforming the protonmotive force into mechanical rotation. The MotA protein is a proton channel. Morphological features such as rings, rod etc. are indicated. The S and M rings probably constitute the rotor; the switch complex controls the rotational direction. The filament is relatively much longer (about 10 μm) than shown here. A channel through which the extracellular protein subunits travel to reach their site of assembly is shown in dotted outline. HAP, hook associated protein. The stator and motor components have not been firmly identified (McNab, 1990).

repulsion from positive charges on a stator component. The latter must be firmly anchored, probably by the peptidoglycan layer in the cell wall. At any given protonmotive force the higher the number of protons translocated per rotation, the greater will be the torque. This proton stoichiometry is not known with certainty but calculated values are in the range 300–1000, with the efficiency varying from close to 100% at high load to 10% at low loads.

The flagella do not rotate continuously. The rotational movement of the external filament can be interrupted (or briefly stopped) and its direction reversed (i.e. clockwise or counterclockwise). The direction and duration of rotation are controlled by a sensory system so that the direction of swimming is biased towards nutrients but away from toxic molecules and other disadvantageous environments. The mechanism of this control is beyond the scope of this book (see Armitage, 1992).

8.7 TRANSPORT OF MACROMOLECULES ACROSS MEMBRANES

Reviews   Pfanner and Neupert 1990, Wickner *et al.* 1991

Bacteria synthesize all of their proteins in the cytoplasm but the destination of some of the proteins may be the periplasm, the outer membrane or even the external medium. In these cases the newly synthesized polypeptide must be transported across the cytoplasmic membrane. Such transport appears in general to require firstly the participation of the *sec* gene products, secondly the retention by the polypeptides of a relatively unfolded state for which ATP may be required, and thirdly a leader or signal sequence of approximately 20–25 amino acids at the N-terminus of the polypeptide. The signal sequence is not conserved between proteins but some features are always found, including the presence of several positively charged residues at the N-terminus and a hydrophobic sequence in the middle.

The connection with the subject matter of this book is that there is good evidence that Δp is involved in driving the translocation of nascent polypeptides across the cytoplasmic membrane. How this is done remains a matter of conjecture; a general electrophoretic effect is improbable because the charge on the leader sequence is of the wrong polarity and both $\Delta\psi$ and ΔpH are competent. It has been suggested that the positive N-terminal end of the leader sequence may be anchored to the cytoplasmic side of the membrane by the $\Delta\psi$; however, it is evident that the membrane potential only acts across the membrane dielectric. Molecules that are confined to one side of the membrane are not affected by the membrane potential. Thus any such anchoring of the positively charged N-terminus of the leader sequence would probably have to be driven indirectly via an effect of membrane potential on a transmembrane protein.

Perhaps a more plausible possibility is that one of the components of the *sec* system, the Sec Y/E protein, may act as a proton/polypeptide antiport system. In this context protein movement within the secretory apparatus could be the basis of the dependence on Δp. In simple terms we could imagine an exaggerated 'breathing' movement of a channel-type protein in the membrane being responsible for squeezing the polypeptide across the membrane. The precedent for the mechanical movement of proteins driven by the proton-motive force is set by the flagella system (Section 8.6).

The import of proteins into mitochondria poses related problems. ATP is again required to maintain an import-competent state, but in this case it is known that the N-terminal targeting sequences are distinct from those for bacterial targeting. They tend to have a predominance of positive charge throughout and many could form amphiphilic helices with the positive charge on one side. $\Delta \psi$, but not ΔpH – a rare example of the lack of interchangeability of the two components of Δp – is required for import into the mitochondrion and the negative polarity of the matrix phase has again prompted consideration of an electrophoretic mechanism. For several reasons, including those discussed above, and the high energy barrier which would be involved in electrophoresizing polypeptides through a membrane, this seems unlikely. It does seem clear that it is the insertion into the inner mitochondrial membrane and translocation of the targeting sequence that is dependent upon $\Delta \psi$ and that the remaining part of the protein can follow independently. These observations may indicate that it is a receptor that is acted upon by the $\Delta \psi$ but full understanding of the role of membrane potential in targeting proteins across the inner mitochondrial membrane is likely to require a full characterization and reconstitution of the proteinaceous components involved. Elsewhere in this book we have seen that such knowledge is required before mechanistic details of the role of Δp in driving endergonic reactions can be appreciated.

9 ENDPIECE

CONNECTIONS TO MEMBRANES IN GENERAL

As we explained in the introductory Chapter 1, this book is concerned with a coherent body of knowledge relating to energy transduction processes associated with three types of membranes: inner mitochondrial, bacterial cytoplasmic and thylakoid. It is, however, important to realize that many of the principles discussed in this book apply to the activities of other types of membranes found in cells. For example, there are proton pumping ATPases found in plasma membranes of lower eukaryotes and vacuolar membranes of various cell types including plants. Each type of ATPase generates a proton electrochemical gradient and transport processes occur by proton symports or antiports and the related processes discussed in this book. The finding of sodium symporters for some transport processes in bacteria also has its clear parallel with such processes as glucose and amino acid sodium-symport across the plasma membrane of some eukaryotic cells. The latter membranes do not, however, contain a proton pumping ATPase but rather the Na/K ATPase that is ubiquitous in higher eukaryotes. As a further example, the bioenergetics of neurotransmitter release can be interpreted by developing the ideas of trans-membrane ion fluxes developed in this book (e.g., McMahon and Nicholls, 1991). By way of illustration we consider briefly two specific systems that illustrate the applications to other membrane systems of the concepts and methodologies discussed in this book.

Whereas we emphasized in Chapter 5 that cytochrome oxidase of energy transducing membranes must not produce superoxide ($O_2^{\cdot-}$), the plasma membrane electron transport systems of phagocytic cells are, on the contrary, designed to produce this reactive radical as a defensive weapon (Cross and Jones, 1991, Morel *et al.* 1991). For example, in the neutrophil there is an NADPH oxidase which is normally inactive but can be stimulated by certain signals arriving at the cells to cause a 'respiratory burst'. This oxidase system appears to involve a flavin and a b-type cytochrome of unusually low potential ($E_{m,7} = -245$ mV) that is sufficient to reduce O_2 to $O_2^{\cdot-}$ ($E_{m,7} = -160$ mV). Absence of this cytochrome causes granulomatous disease. Electron transfer from cytosolic NADPH to the site of $O_2^{\cdot-}$ production at the external surface of the cells could generate a membrane potential, negative outside, that would build up sufficiently after movement of a small amount of charge (see Chapter 3) to prevent further $O_2^{\cdot-}$ production. It is proposed that the outward movement of electrons is charge compensated by the movement of cations, probably H^+, in the same direction (Cross and Jones, 1991).

The chromaffin vesicles (formerly called granules) of adrenal medulla serve as a second example (Njus *et al.*, 1986). These vesicles store the catecholamine adrenaline (epinephrine). The membranes of the vesicles contain an ATPase, now known to be a representative example of the V-class of proton translocating ATPases (Nelson, 1989). This enzyme hydrolyses cytosolic ATP to pump protons from the cytoplasm to the interior of the vesicles which is both more acidic and positively charged (i.e. a P-phase) than the cytoplasm. Catecholamines are weak bases, and in principle the uptake could be via passive equilibration of the uncharged weak base form by exactly the mode discussed for weak bases in Chapter 4. However, it turns out that such a mechanism would not account for the magnitude of the accumulation ratio and the uptake is mediated by a catecholamine/proton antiporter that is inhibited by reserpine. Either the protonated catecholamine moves in antiport for two protons (thus the driving force is dependent on $\Delta\psi$ plus 2 ΔpH) or the uncharged catecholamine moves in antiport for one proton; these modes are difficult to distinguish experimentally. The granules have a high internal concentration of ATP; the uptake mechanism is unknown but may be driven by the proton electrochemical gradient across the membrane (Njus *et al.*, 1986). The membranes of the granules also contain a cytochrome b_{561} ($E_{m,7} = 140$ mV) which serves to transfer electrons from the cytoplasm to internal semidehydroascorbate, thus forming the ascorbate required for dopamine beta-hydroxylase which generates noradrenaline from dopamine with concomitant formation of semidehydroascorbate. The cytoplasmic electron source is ascorbate, which is oxidized in a one electron process to semidehydroascorbate by the cytochrome b_{561}; ascorbate is subsequently regenerated by reduction of the semidehydroascorbate with NADH. The transfer of electrons across the membrane is driven by the membrane potential, positive inside, that is maintained by the activity of the proton pumping ATPase, and also by the pH gradient because the pH-dependency (Chapter 3) of the ascorbate/semidehydrosascorbate couple is such that formation of internal ascorbate is favoured (Njus *et al.*, 1986). Thus chemiosmotic energy transduction principles and methods for characterizing electron transfer proteins and proton translocating ATPases are demonstrably relevant to the functioning of the chromaffin vesicle, just as they are to other types of similar vesicles, for instance those containing serotonin (5-hydroxytryptamine) or acetylcholine (Njus *et al.*, 1986). The messages of this book are clearly not just applicable to the bioenergetics of mitochondria, bacteria and chloroplasts!

REFERENCES

Allen, J.F. (1992) How does protein phosphorylation regulate photosynthesis? *Trends Biochem. Sci.* **17**, 12–17.

Ambudkar, S.V. and Rosen, B.P. (1990) Ion-exchange systems in prokaryotes. In: *The Bacteria: Bacterial Energetics* (T.A. Krulwich, ed.), Academic Press, San Diego, vol. XII, pp. 247–271.

Ames, G.F-L. (1990) Energetics of periplasmic transport systems. In: *The Bacteria: Bacterial Energetics* (T.A. Krulwich, ed.), Academic Press, San Diego, vol. XII, pp. 225–246.

Andersson, B. and Styring, S. (1991) Photosystem II: Molecular organization, function, and acclimation. *Curr. Top. Bioenerget.* **16**, 1–81.

Anraku, Y. (1988) Bacterial electron transport chains. *Annu. Rev. Biochem.* **57**, 101–132.

Anraku Y. and Gennis R.B. (1987) The aerobic respiratory chain of Escherichia coli. *Trends. Biochem. Sci.* **12**, 262–266.

Anthony, C.A. (1988) *Bacterial Energy Transduction*. Academic Press, London, pp. 1–517.

Armitage, J.P. (1992) Behavioral responses in bacteria. *Annu. Rev. Physiol.* **54**, 683–714.

Azzone, G.F., Zoratti M., Petronilli V. and Pietrobon D. (1985) The stoichiometry of H^+ pumping in cytochrome oxidase and the mechanism of uncoupling. *J. Inorg. Biochem.* **23**, 349–356.

Babcock, G.T. and Wikström, M. (1992) Oxygen activation and the conservation of energy in cell respiration. *Nature* **356**, 301–309.

Baltscheffsky, M. and Baltscheffsky, H. (1992) Inorganic pyrophosphate and inorganic pyrophosphatases. In: *New Comprehensive Biochemistry*. (L. Ernster, ed.), Elsevier, Amsterdam, in press.

Bianchet, M., Ysern, Y., Hullihen, J., Pedersen, P.L. and Amzel, L.M. (1991) Mitochondrial ATP synthase – quaternary structure of the F_1 moiety at 3.6 Å resolution determined by X-ray diffraction analysis. *J. Biol. Chem.* **266**, 21197–21201.

Blaut, M., Muller, V. and Gottschalk, G. (1990) Energetics of methanogens. In: *The Bacteria: Bacterial Energetics* (T.A. Krulwich, ed.), Academic Press, San Diego and London, vol. XII, pp. 505–537.

Booth, I.R. (1988) Bacterial transport: energetics and mechanisms. In: *Bacterial Energy Transduction* (C. Anthony, ed.), Academic Press, London, pp. 377–428.

Boussac, A., Zimmermann, J.L., Rutherford, A.W. and Lavergne J. (1990) Histidine oxidation in the oxygen-evolving photosystem II enzyme. *Nature* **347**, 303–306.

Boyer P.D. (1989) A perspective of the binding change mechanism for ATP synthesis. *FASEB. J.* **3**, 2164–2178.

Brand, M.D. and Murphy M.P. (1987) Control of electron flux through the respiratory chain in mitochondria and cells. *Biol. Rev.* **62**, 141–193.

Brooker, R.J. (1990) The lactose permease of Escherichia coli. *Res. Microbiol.* **141**, 309–315.

Capaldi, R.A. (1988) Mitochondrial myopathies and respiratory chain proteins. *Trends. Biochem. Sci.* **13**, 144–148.

Capaldi, R. (1990) Structure and function of cytochrome c oxidase. *Annu. Rev. Biochem.* **59**, 569–596.

Chan, S.I. and Li, P.M. (1990) Cytochrome c oxidase: understanding nature's design of a proton pump. *Biochemistry* **29**, 1–12.

Chance, B. and Williams, G.R. (1956) The respiratory chain and oxidative phosphorylation. *Adv. Enzymol.* **17**, 65–134.

Chappell, J.B. (1968) Systems used for the transport of substrates into mitochondria. *Br. Med. Bull.* **24**, 150–157.

Cooper, J.M., Schapira, A.H.V., Holt, I.J., Toscano, A., Harding, A.E., Morgan-Hughes, J.A. and Clark, J.B. (1991) Biochemical and molecular aspects of human mitochondrial respiratory chain disorders. *Biochem. Soc. Trans.* **18**, 517–519.

Cornish-Bowden, A. (1983) Metabolic efficiency: is it a useful concept?. *Biochem. Soc. Trans.* **11**, 44–45.

Cramer, W.A. and Knaff, D.B. (1989) *Energy Transduction in Biological Membranes.* Springer-Verlag, New York and Heidelberg.

Cross, A.R., and Jones, O.T.G. (1991) Enzymic mechanisms of superoxide production. *Biochim. Biophys. Acta* **1057**, 281–298.

Davis, E.J. and Davis van Thienen, W.I. (1989) Force-flow and back-pressure relationships in mitochondrial energy transduction: an examination of extended state 3–state 4 transitions. *Arch. Biochem. Biophys.* **275**, 449–458.

Deisenhofer, J. and Michel, H. (1991) High-resolution structures of photosynthetic reaction centers. *Annu. Rev. Biophys. Biophys. Chem.* **20**, 247–266.

Dibrov, P.A. (1991) The role of sodium ion transport in *Escherichia coli* energetics. *Biochim. Biophys. Acta* **1056**, 209–224.

DiMarco, A.A., Bobik, T.A. and Wolfe, R.S. (1990) Unusual coenzymes of methanogenesis. *Annu. Rev. Biochem.* **59**, 355–394.

Dimroth, P. (1991) Na^+-coupled alternative to H^+ coupled primary transport system in bacteria. *Bioessays* **13**, 463–468.

Douce, R. and Neuburger, M. (1989) The uniqueness of plant mitochondria. *Annu. Rev. Plant. Physiol.* **40**, 371–414.

Duine, J. A. (1991) Quinoproteins: enzymes containing the quinone cofactor pyrroloquinoline quinone, topaquinone or tryptophan tryptophan quinone. *Eur. J. Biochem.* **200**, 271–284.

Duszynski, J. and Wojtczak, L. (1985) The apparent non-linearity of the relationship between the rate of respiration and the protonmotive force of mitochondria can be explained by heterogeneity of mitochondrial preparations. *FEBS Lett.* **182**, 243–248.

Dutton, P.L. (1978) Redox potentiometry: determination of midpoint potentials of oxidation–reduction components of biological electron transport systems. *Methods Enzymol.* **54**, 411–435.

Epstein, W. (1990) Bacterial transport ATPases. In: *The Bacteria: Bacterial Energetics* (T.A. Krulwich, ed.), Academic Press, San Diego, vol. XII, pp. 87–110.

Feher, G., Allen, J.P., Okamura, M.Y. and Rees, D.C. (1989) Structure and function of bacterial photosynthetic reaction centres. *Nature* **331**, 111–116.

Ferguson, S.J. (1985) Fully delocalized chemiosmotic coupling or localized proton flow pathways in energy coupling? A scrutiny of experimental evidence. *Biochim. Biophys. Acta* **811**, 47–95.

Ferguson, S.J. (1986) The ups and downs of P/O ratios (and the question of non-integral coupling stoichiometries for oxidative phosphorylation and related processes. *Trends. Biochem. Sci.* **11**, 351–353.

Ferguson, S.J. (1987a) The redox reactions of the nitrogen and sulphur cycles. *Symp. Soc. Gen. Microbiol.* **42**, 1–29.

Ferguson, S.J. (1987b) Denitrification: a question of the control and organization of electron and ion transport. *Trends. Biochem. Sci.* **12**, 354–357.

Ferguson, S.J. (1992) The periplasm. *Symp. Soc. Gen. Microbiol.* **47**, 297–325.

Ferguson, S.J., Jackson J.B. and McEwan, A.G. (1987) Anaerobic respiration in the Rhodospirillaceae: characterisation of pathways and evaluation of roles in redox balancing during photosynthesis. *FEMS. Microbiol. Lett.* **46**, 117–143.

Ferreira, G.C., Pratt, R.D. and Pedersen, P.L. (1989) Energy-linked anion transport. Cloning, sequencing, and characterization of a full length cDNA encoding the rat liver mitochondrial proton/phosphate symporter. *J. Biol. Chem.* **264**, 15628–15633.

Fillingame, R.H. (1990) Molecular mechanics of ATP synthesis of F_1F_o-type H^+-transporting ATP synthases. In: *The Bacteria: Bacterial Energetics* (T.A. Krulwich, ed.), Academic Press, San Diego, vol. XII, pp. 345–391.

Franco, P.J. and Brooker, R.J. (1991) Evidence that the asparagine 322 mutant of the lactose permease transports protons and lactose with a normal stoichiometry and accumulates lactose against a concentration gradient. *J. Biol. Chem.* **266**, 6693–6699.

Garlid, K.D. (1988) Sodium/proton antiporters in the mitochondrial inner membrane. *Adv. Exp. Med. Biol.* **232**, 37–46.

Gilbert, G.N. and Mulkay, M. (1984) *Opening Pandora's Box.* Cambridge University Press, Cambridge. pp. 202.

Godden, J.W., Turley, S., Teller, D.C., Adman, E.T., Liu, M.Y., Payne, W.J. and LeGall, J. (1991) The 2.3 Å X-ray structure of nitrite reductase from Achromobacter cycloclastes. *Science* **253**, 438–442.

Gogol, E.P., Lucken, U. and Capaldi, R.A. (1987) The stalk connecting the F1 and Fo domains of ATP synthase visualized by electron microscopy of unstained specimens. *FEBS Lett.* **219**, 274–278.

Golbeck, J.H. and Bryant, D.A. (1989) Photosystem I. *Curr. Top. Bioenerget.* **16**, 83–177.

Gottschalk, G. (1986) *Bacterial Metabolism*. 2nd edn, Springer-Verlag, New York, pp. 359.

Gregory, R.P.F. (1989) *Biochemistry of Photosynthesis*. Wiley, Chichester, pp. 290.

Grivell, L.A. and Jacobs, H.T. (1991) Oncogenes, mitochondria and immortality. *Current Biology*. **1**, 94–96.

Groen, A.K., Wanders, R.J.A., Westerhoff, H.V., van der Meer, R. and Tager, J.M. (1983) Quantification of the contribution of various steps to the control of mitochondrial respiration. *J. Biol. Chem.* **257**, 2754–2757.

Guilfoile, P.G. and Hutchinson, C.R. (1991) A bacterial analog of the mdr gene of mammalian tumor cells is present in *Streptomyces peucetius*, the producer of daunorubicin and doxorubicin. *Proc. Natl. Acad. Sci. USA* **88**, 8553–8557.

Hafner, R.P., Brown, G.C. and Brand, M.D. (1990) Analysis of the control of respiration rate, phosphorylation rate, proton leak rate and protonmotive force in isolated mitochondria using the 'top-down' approach of metabolic control theory. *Eur. J. Biochem.* **188**, 313–319.

Haltia, T., Saraste, M. and Wikstrom, M. (1991) Subunit III of cytochrome c oxidase is not involved in proton translocation: a site-directed mutagenesis study. *EMBO J.* **10**, 2015–2021.

Harris, D.A. and Das, A.M. (1991) Control of mitochondrial ATP synthesis in the heart. *Biochem. J.* **280**, 561–573.

Henderson P.J.F. (1990) Proton-linked sugar transport systems in bacteria. *J. Bioenerg. Biomemb.* **22**, 525–569.

Henderson, P.J.F. (1991) Sugar transport proteins. *Curr. Opinion Struct. Biol.* **1**, 590–601.

Henderson, R., Baldwin, J.M., Ceska, T.A., Zemlin, T.A., Beckman, E. and Downing, K.H. (1990) Model for the structure of bacteriorhodopsin based on high resolution electron cryo-microscopy. *J. Mol. Biol.* **213**, 899–929.

Higgins, C.F. (1992) ABC transporters: from microorganisms to man. *Annu. Rev. Cell Biol.* **8** in press.

Higgins, C.F., Hyde, S.C., Mimmack, M.M., Gileadi, U., Gill, D.R. and Gallagher, M.P. (1990) Binding protein-dependent transport systems. *J. Bioenerg. Biomemb.* **22**, 571–592.

Hinkle, P.C., Kumar, M.A., Resetar, A. and Harris, D.L. (1991) Mechanistic stoichiometry of mitochondrial oxidative phosphorylation. *Biochemistry* **30**, 3576–3582.

Hirata, H., Ohno, K., Sone, N., Kagawa, Y. and Hamamoto, T. (1986) Direct measurement of the electrogenicity of the H^+-ATPase from thermophilic bacterium PS3 reconstituted in planar phospholipid bilayers. *J. Biol. Chem.* **251**, 9839–9843.

Hockenbery, D., Nunez, G., Milliman, C., Schreiber, R. D. and Korsmeyer, S.J. (1990) Bcl-2 is an inner mitochondrial membrane protein that blocks programmed cell death. *Nature* **348**, 334–336.

Hofmann, A. and Dimroth, P. (1991) The electrochemical proton potential of *Bacillus alcalophilus*. *Eur. J. Bichoem.* **201**, 467–473.

Holzapfel, W., Finkele, U., Kaiser, W., Oesterhelt, D., Scheer, H., Stilz, H.U. and Zinth, W. (1990) Initial electron-transfer in the reaction center from *Rhodobacter sphaeroides*. *Proc. Natl. Acad. Sci. USA* **87**, 5168–5172.

Hunter, C.N., van Grondelle, R. and Olsen, J.D. (1989) Photosynthetic antenna proteins; 100 psecs before photochemistry starts. *Trends. Biochem. Sci.* **14**, 72–76.

Ingledew, W.J. (1982) The bioenergetics of an acidophilic chemolithotroph. *Biochim. Biophys. Acta* **683**, 89–117.

Jackson, J.B. (1988) Bacterial photosynthesis. In: *Bacterial Energy Transduction* (C. Anthony, ed.), Academic Press, London, pp. 317–375.

Jackson, J.B. (1991) The proton translocating nicotinamide adenine dinucleotide transhydrogenase. *J. Bioenerg. Biomemb.* **23**, 715–741.

Jagendorf, A.T. and Uribe, E. (1966) ATP formation caused by acid–base transition of spinach chloroplasts. *Proc. Natl. Acad. Sci. USA* **55**, 170–177.

Jezek, P., Mahdi, F. and Garlid, K.D. (1990) Reconstitution of the beef heart and rat liver mitochondrial K^+/H^+ (Na^+/H^+) antiporter. Quantitation of K^+ transport with the novel fluorescent probe, PBFI. *J. Biol. Chem.* **265**, 10522–10526.

Kaback, H.R. (1990) Lac permease of *Escherichia coli*: on the path of the proton. *Phil. Trans. R. Soc. Lond.* **B326**, 425–436.

Kaback, H.R., Bibi, E. and Roepe, P.D. (1990) β-Galactosidase transport in *E. coli*: a functional dissection of lac permease. *Trends Biochem. Sci.* **15**, 309–314.

Kagawa, Y., Kandrach, A. and Racker, E. (1973) Partial resolution of the enzymes catalysing oxidative phosphorylation. *J. Biol. Chem.* **248**, 676–684.

Khorana, H.G. (1988) Bacteriorhodopsin, a membrane protein that uses light to translocate protons. *J. Biol. Chem.* **263**, 7439–7442.

Kirmaier, C. and Holten, D. (1991) An assessment of the mechanism of initial electron transfer in bacterial reaction centres. *Biochemistry* **30**, 609–613.

Klingenberg, M. (1989) Molecular aspects of the adenine nucleotide carrier from mitochondria. *Arch. Biochem. Biophys.* **270**, 1–14.

Klingenberg, M. (1990) Mechanism and evolution of the uncoupling protein of brown adipose tissue. *Trends. Biochem. Sci.* **15**, 108–112.

Koch, M.H.J., Oesterhelt, D., Plohn, H.J., Rapp, G. and Buldt, G. (1991) Time-resolved X-ray diffraction study of structural changes associated with the photocyle of bacteriorhodopsin. *EMBO J.* **10**, 521–526.

Kornberg, H.L. and Henderson, P.J.F. (1990) *Microbial Membrane Transport Systems*. The Royal Society, London.

Krishnamoorthy, G. and Hinkle, P.C. (1984) Non-ohmic proton conductance of mitochondria and liposomes. *Biochemistry* **23**, 1640–1645.

Krulwich, T.A., Hicks, D.B., Seto-Young, D. and Guffanti, A.A. (1988) The bioenergetics of alkalophilic bacteria. *CRC Crit. Rev. Microbiol.* **16**, 15–36.

Kuhlbrandt, W. and Da Neng Wang (1991) Three-dimensional structure of plant light-harvesting complex determined by electron crystallography. *Nature* **350**, 130–134.

Lehninger, A.L., Reynafarje, B., Alexandre, A. and Villalobo, A. (1979) Proton stoichiometry and mechanisms in mitochondrial energy transduction. In: *Membrane Bioenergetics* (C.P. Lee, ed.), Addison-Wesley, London, pp. 393–404.

Lengeler, J.W. (1990) Molecular analysis of the Enzyme II-complexes of the bacterial phosphotransferase system (PTS) as carbohydrate transport systems. *Biochim. Biophys. Acta* **1018**, 155–159.

Levings, C.S. (1990) The Texas cytoplasm of maize: cytoplasmic male sterility and disease susceptibility. *Science* **250**, 942–947.

Lin, E.C.C. and Kuritzkes, D.R. (1987) Pathways for anaerobic respiration. In: *Escherichia coli and Salmonella typhimurium: Cellular and Molecular Biology* (F.C. Neidehardt, J.L. Ingraham, K.B. Low, B. Magasanik, M. Schaechter and H.E. Umberger, eds), *Am. Soc. Microbiol.*, Washington, D.C.; pp. 201–221.

Locke, R.M., Rial, E., Scott, I.D. and Nicholls, D.G. (1982) Fatty acids as acute regulators of the proton conductance of hamster brown fat mitochondria. *Eur. J. Biochem.* **129**, 373–380.

Malmstrom, B.G. (1990) Cytochrome oxidase: some unsolved problems and controversial issues. *Arch. Biochem. Biophys.* **280**, 233–241.

Maloney, P.C. (1990) Anion exchange reactions in bacteria. *J. Bioenerg. Biomemb.* **22**, 509–523.

Maloy, S.R. (1990) Sodium-coupled cotransport. In: *The Bacteria: Bacterial Energetics* (T.A. Krulwich, ed.), Academic Press, San Diego, vol. XII, pp. 203–224.

Mathies, R.A., Lin, S.W., Ames, J.B. and Pollard, T.W. (1991) From femtoseconds to biology: mechanism of bacteriorhodopsin's light-driven proton pump. *Annu. Rev. Biophys. Chem.* **20**, 491–518.

Matthews, F.S., Chen, L. and Durley, R. (1991) Crystal structure of an electron-transfer complex between a quinoprotein and a blue copper protein: methylamine dehydrogenase and amicyanin. *J. Inorg. Biochem.* **43**, 101.

McCormack, J.G., Halestrap, A.P. and Denton, R.M. (1990) Role of calcium ions in regulation of mammalian intramitochondrial metabolism. *Physiol. Rev.* **70**, 391–425.

McIntire, W.S., Wemmer, D.E., Christoserdov, M. and Lidstrom, M.E. (1991) A new cofactor in prokaryotic enzyme: tryptophan tryptophanylquinone as the redox prosthetic group in methylamine dehydrogenase. *Science* **252**, 817–824.

McMahon, H.T. and Nicholls, D.G. (1991) The bioenergetics of neurotransmitter release *Biochim. Biophys. Acta* **1059**, 243–264.

McNab, R.M. (1990) The genetics, structure and assembly of the bacterial flagellum. *Symp. Soc. Gen. Microbiol.* **46**, 77–106.

Meadow, N.D., Fox, D.K. and Roseman, S. (1991) The bacterial phosphoenol-pyruvate:glycose phospotransferase system. *Annu. Rev. Biochem.* **59**, 497–542.

Mitchell, P. and Moyle, J. (1965) Stoichiometry of proton translocation through the respiratory chain and adenosine triphosphatase systems of rat liver mitochondria. *Nature* **208**, 147–151.

Mitchell, P. and Moyle, J. (1967a) Respiration-driven proton translocation in rat liver mitochondria. *Biochem. J.* **105**, 1147–1162.

Mitchell, P. and Moyle, J. (1967b) Acid–base titration across the membrane system of rat-liver mitochondria. *Biochem. J.* **104**, 588–600.

Mitchell, P. and Moyle, J. (1969) Estimation of membrane potential and pH difference across the cristal membrane of rat liver mitochondria. *Eur. J. Biochem.* **7**, 471–478.

Moore, A.L. and Siedow, J.N. (1991) The regulation and nature of the cyanide-resistant alternative oxidase of plant mitochondria. *Biochim. Biophys. Acta* **1059**, 121–140.

Moore, G.R. and Pettigrew, G.W. (1990) *Cytochromes c: Evolutionary, Structural and Physiochemical Aspects*. Springer-Verlag, Heidelberg.

Morel, F., Doussiere, J. and Vignais, P.V. (1991) The superoxide generating oxidase b_{558} of phagocytic cells; physiological, molecular, and pathological aspects. *Eur. J. Biochem.* **201**, 523–546.

Morgan-Hughes, J.A., Schapira, A.H.V., Cooper, J.M., Holt, I.J., Harding, A.E. and Clark, J.B. (1990) The molecular pathology of respiratory-chain dysfunction in human mitochondrial myopathies. *Biochim. Biophys. Acta* **1018**, 217–222.

Moser, C.C., Keske, J.M., Warncke, K., Farid, R.S. and Dutton, P.L. (1992) Nature of biological electron transfer. *Nature* **355**, 796–802.

Murphy, M.P. and Brand, M.D. (1987) Variable stoichiometry of proton pumping by the mitochondrial respiratory chain. *Nature* **329**, 170–172.

Murphy, M.P. and Brand, M.D. (1988) Membrane-potential-dependent changes in the stoichiometry of charge translocation by the mitochondrial electron transport chain. *Eur. J. Biochem.* **173**, 637–644.

Nelson, N. (1989) Structure, molecular genetics and evolution of vacuolar H$^+$ ATPases. *J. Bioenerg. Biomemb.* **21**, 553–571.

Neumann, J. and Jagendorf, A.T. (1964) Light-induced pH changes related to phosphorylation by chloroplasts. *Arch. Biochem. Biophys.* **107**, 109–119.

Nicholls, D.G. (1974a) The influence of respiration and ATP hydrolysis of the proton electrochemical potential gradient across the inner membrane of rat liver mitochondria as determined by ion distribution. *Eur. J. Biochem.* **50**, 305–315.

Nicholls, D.G. (1974b) Hamster brown adipose tissue mitochondria: the control of respiration and the proton electrochemical potential by possible physiological effectors of the proton conductance of the inner membrane. *Eur. J. Biochem.* **49**, 573–583.

Nicholls, D.G. and Åkerman, K.E.O. (1982) Mitochondrial Ca transport. *Biochim. Biophys. Acta* **683**, 57–88.

Nicholls, D.G. and Attwell, D. (1990) Excitatory amino acid pharmacology: the release and reuptake of excitatory amino acids. *Trends Pharmacol. Sci.* **11**, 462–468.

Nicholls, D.G. and Bernson, V.S.M. (1977) Inter-relationships between proton electrochemical gradient, adenine nucleotide phosphorylation potential and respiration during substrate-level and oxidative phosphorylation by mitochondria from brown adipose tissue of cold-adapted guinea-pigs. *Eur. J. Biochem.* **75**, 601–612.

Nicholls, D.G. and Locke, R.M. (1984) Thermogenic mechanisms in brown fat. *Physiol. Rev.* **64**, 1–64.

Nitschke, W. and Rutherford, A.W. (1991) Photosynthetic reaction centres: variations on a common structural theme. *Trends Biochem. Sci.* **16**, 241–245.

Njus, D., Kelley, P.M. and Harnadek, G.J. (1986) Bioenergetics of secretory vesicles. *Biochim. Biophys. Acta* **853**, 237–265.

Nobes, C.D., Brown, G.C., Olive, P.N. and Brand, M.D. (1990) Non-ohmic proton conductance of the mitochondrial inner membrane in hepatocytes. *J. Biol. Chem.* **265**, 12903–12909.

Oesterhelt, D. and Tittor, J. (1989) Two pumps, one principle: light-driven transport in Halobacteria. *Trends Biochem. Sci.* **14**, 57–61.

Paddock, M.L., Rongey, S.H., Feher, G. and Okamura, M.Y. (1989) Pathway of proton transfer in bacterial reaction centers: replacement of glutamic acid 212 in the L subunit by glutamine inhibits quinone (secondary acceptor) turnover. *Proc. Natl. Acad. Sci. USA* **86**, 6602–6606.

Papiz, M.Z., Hawthornthwaite, A.M., Cogdell, R.J., Woolley, K.J., Wightman, P.A., Ferguson, L.A. and Lindsay, J.G. (1989) Crystallization and characterization of two crystal forms of the B800-850 light-harvesting complex from *Rhodopseudomonas acidophila* strain 10050. *J. Mol. Biol.* **209**, 833–835.

Penefsky, H.S. and Cross, R.L. (1991) Structure and mechanism of FoF1-type ATP synthases and ATPases. *Adv. Enzymol.* **64**, 173–214.

Perez J.A. and Ferguson, S.J. (1990) Kinetics of oxidative phosphorylation in *Paracoccus denitrificans*. 1. Mechanism of ATP synthesis at the active site(s) of F_0F_1 ATPase. *Biochemistry* **29**, 10503–10518.

Pettigrew, G.W. and Moore, G.R. (1987) *Cytochromes c: Biological Aspects.* Springer-Verlag, Heidelberg.

Pfanner, N. and Neupert, W. (1990) The mitochondrial protein import apparatus. *Annu. Rev. Biochem.* **59**, 331–353.

Pietrobon, D., Zoratti, M., Azzone, G.F. and Caplan, S.R. (1986) Intrinsic uncoupling of mitochondrial proton pumps. 2. Modeling studies. *Biochemistry* **25**, 767–775.

Poole, R.K. and Ingledew, W.J. (1987) Pathways of electrons to oxygen. In: *Escherichia coli and Salmonella typhimurium: Cellular and Molecular Biology* (F.C. Neidehardt, J.L. Ingraham, K.B. Low, B. Magasanik, M. Schaechter and H.E. Umberger, eds), *Am. Soc. Micriobiol.*, Washington D.C., pp. 170–200.

Prince, R.C. and George, G.N. (1990) Tryptophan radicals. *Trends Biochem. Sci.* **15**, 170–171.

Prince, R.C., Linkletter, S.J.G. and Dutton, P.L. (1981) The thermodynamic properties of some commonly used oxidation–reduction mediators, inhibitors and dyes, as determined by polarography. *Biochim. Biophys. Acta* **635**, 132–148.

Puustinen, A., Finel, M., Haltia, T., Gennis, R.B. and Wikstrom, M. (1991) Properties of the two terminal oxidases of *Escherichia coli*. *Biochemistry* **30**, 3936–3942.

Ramsay, R.R., Dadgar, J., Trevor, A. and Singer, T.P. (1986) Energy-driven uptake of N-methyl-4-phenylpyridine by brain mitochondria mediates the neurotoxicity of MPTP. *Life. Sci.* **39**, 581–588.

Rottenberg, H. (1986) Fatty acid uncoupling of oxidative phosphorylation in rat liver mitochondria. *Biochemistry* **25**, 1747–1755.

Runswick, M.J., Walker, J.E., Biscaccia, F., Icacobazzi, V. and Palmieri, F. (1990) Sequence of the bovine 2-oxoglutarate/malate carrier protein: structural relationship to other mitochondrial transport proteins. *Biochemistry* **29**, 11033–11040.

Rutherford, A.W. (1989) Photosystem II, the water-splitting enzyme. *Trends Biochem. Sci.* **14**, 227–232.

Saraste, M. (1990) Structural features of cytochrome oxidase. *Q. Rev. Biophys.* **23**, 331–366.

Schneider, E. and Altendorf, K. (1987) Bacterial adenosine 5′-triphosphate synthase (F_1F_0): purification and reconstitution of Fo complexes and biochemical and functional characterization of their subunits. *Microbiol. Rev.* **51**, 477–497.

Schonfeld, P., Schild, L. and Kunz, W. (1989) Long-chain fatty acids act as protonophoric uncouplers of oxidative phosphorylation in rat liver mitochondria. *Biochim. Biophys. Acta* **977**, 266–272.

Selman-Reimer, S., Duhe, R.J., Stockman, B.J. and Selman, B.R. (1991) L-1-*N*-methyl-4-mercaptohistidine disulfide, a potent endogenous regulator in the redox control of chloroplast coupling factor 1 in *Dunaliella*. *J. Biol. Chem.* **266**, 182–188.

Senior, A.E. (1990) The proton-translocating ATPase of *Escherichia coli*. *Annu. Rev. Biophys. Chem.* **19**, 7–41.

Skulachev, V.P. (1970) Electric fields in coupling membranes. *FEBS Lett.* **11**, 301–319.

Skulachev, V.P. (1988) *Membrane Bioenergetics*. Springer-Verlag, Berlin, 442 pp.

Sorgato, M., Lippe, G., Seren, S. and Ferguson, S.J. (1985) Partial uncoupling, or inhibition of electron transport rate, have equivalent effects on the relationship between the rate of ATP synthesis and proton-motive force in submitochondrial particles. *FEBS Lett.* **181**, 323–327.

Spiro, S. and Guest, J.R. (1991) Adaptive responses to oxygen limitation in *Escherichia coli*. *Trends Biochem. Sci.* **16**, 310–314.

Stucki, J.W. (1983) Thermodynamic optimization of biological energy conversions. *Biochem. Soc. Trans.* **11**, 45–47.

Takahashi, E. and Wraight, C.A. (1992) Proton and electron transfer in the acceptor quinone complex of *Rhodobacter sphaeroides* reaction center: characterization of site-directed mutants of the two ionizable residues, Glu^{212} and Asp^{213}, in the Q_B binding site. *Biochemistry* **31**, 855–866.

Thauer, R.K. (1990) Energy metabolism of methanogenic bacteria. *Biochim. Biophys. Acta* **1018**, 256–259.

Trumpower, B.L. (1990a) The protonmotive Q cycle – energy transduction by coupling of proton translocation to electron transfer by the cytochrome bc$_1$ complex. *J. Biol. Chem.* **265**, 11409–11412.

Trumpower, B.L. (1990b) Cytochrome bc1 complexes of microorganisms. *Microbiol. Rev.* **54**, 101–129.

Valverde, M.A., Diaz, M., Sepulveda, F.V., Gill, D.R., Hyde, S.C. and Higgins, C.F. (1992) Volume regulated chloride channel associated with the human multidrug resistance P-glycoprotein. *Nature*, **355**, 830–833.

Varos, G. and Lanyi, J.K. (1991) Thermodynamics and energy coupling in the bacteriorhodopsin photocycle. *Biochemistry* **30**, 5016–5022.

Walker, J.E. (1992) The NADH: ubiquinone oxidoreductase (Complex I) of respiratory chains, *Q. Rev. Biophys.* in press.

Weiss, H., Friedrich, T., Hofhaus, G. and Preis, D. (1991) The respiratory-chain NADH dehydrogenase (complex I) of mitochondria. *Eur. J. Biochem.* **197**, 563–576.

Westerhoff, H.V. and Van Dam, K. (1987) *Thermodynamics and Control of Biological Free-energy Transduction.* Elsevier, Amsterdam and Oxford. pp. 568.

Wickner, W., Driessen, A.J.M. and Hartl, F.U. (1991) The enzymology of protein translocation across the *Escherichia coli* plasma membrane. *Annu. Rev. Biochem.* **60**, 101–124.

Wikström, M. and Babcock, G.T. (1990) Catalytic intermediates. *Nature* **348**, 16–17.

Wittinghofer, A. and Pai, E.F. (1991) The structure of Ras protein: a model for a universal molecular switch. *Trends Biochem. Sci.* **16**, 382–387.

Woelders, H., van der Zande, W.J., Colen, A.M., Wanders, R.J. and Van Dam, K. (1985) The phosphate potential maintained by mitochondria in State 4 is proportional to the proton-motive force. *FEBS Lett.* **179**, 278–282.

Wojtczak, L., Bogucka, K., Duszynski, J., Zablocka, B. and Zolkiewska, A. (1990) Regulation of mitochondrial resting state respiration: slip, leak, heterogeneity? *Biochim. Biophys. Acta* **1018**, 177–181.

Wright, J.K., Seckler, R. and Overath, P. (1986) Molecular aspects of sugar:ion cotransport. *Annu. Rev. Biochem.* **55**, 255–248.

Xu, X., Matsuno-Yagi, A. and Yagi, T. (1991) Characterization of the 25-kilodalton subunit of the energy transducing NADH–ubiquinone oxidoreductase of *Paracoccus denitrificans*: sequence similarity to the 24-kilodalton subunit of the flavoprotein fraction of mammalian complex I. *Biochemistry* **30**, 8678–8684.

Yamanaka, T., Yano, T., Kai, M., Tamegai, H., Sato, A. and Fukumori, Y. (1991) The electron transfer system in an acidophilic iron-oxidising bacterium. In: *New Era of Bioenergetics* (Y. Mukohata, ed.), Academic Press, Tokyo, pp. 223–246.

Zolkiewska, A., Zablocka, B., Duszynski, J. and Wojtczak, L. (1989) Resting state respiration of mitochondria: reappraisal of the role of passive ion fluxes. *Arch. Biochem. Biophys.* **275**, 580–590.

Zoratti, M., Favaron, M., Pietrobon, D. and Azzone, G.F. (1986) Intrinsic uncoupling of mitochondrial proton pumps. 1. Non-ohmic conductance cannot account for the nonlinear dependence of static head respiration on $\Delta\bar{\mu}_{H^+}$. *Biochemistry* **25**, 760–767.

INDEX